MUSHROOMS
HOW TO GROW THEM

A Practical treatise on mushroom culture for profit and pleasure.

MUSHROOMS
HOW TO GROW THEM

A Practical treatise on mushroom culture for profit and pleasure.

By
William Falconer

Illustrated.

MJP PUBLISHERS

CHENNAI   NEW DELHI   TIRUNELVELI

ISBN 978-81-8094-140-5   **MJP PUBLISHERS**
Printed and bound in India   New No. 5, Muthu Kalathy Street
MJP 122   Triplicane, Chennai 600 005
First published: 1892   MJP's first reprint: 2012

BO: 4264/ 3, First Floor, Ansari Road, Daryaganj, New Delhi-2. ☺98961 92710
BO: 70B, Perumal North Car Street, Tirunelveli Junction, Tirunelveli-1 ☺0462- 2330248

PREFACE

Mushrooms and their extensive and profitable culture should concern every one. For home consumption they are a healthful and grateful food, and for market, when successfully grown, they become a most profitable crop. We can have in America the best market in the world for fresh mushrooms; the demand for them is increasing, and the supply has always been inadequate. The price for them here is more than double that paid in any other country, and we have no fear of foreign competition, for all attempts, so far, to import fresh mushrooms from Europe have been unsuccessful.

In the most prosperous and progressive of all countries, with a population of nearly seventy millions of people alert to every profitable, legitimate business, mushroom-growing, one of the simplest and most remunerative of industries, is almost unknown. The market grower already engaged in growing mushrooms appreciates his situation and zealously guards his methods of cultivation from the public. This only incites interest and inquisitiveness, and the people are becoming alive to the fact that there is money in mushrooms and an earnest demand has been created for information about growing them.

The raising of mushrooms is within the reach of nearly every one. Good materials to work with and careful attention to all practical details should give good returns. The industry is one in which women and children can take part as well as men. It furnishes indoor employment in winter, and there is very little hard labor attached to it, while it can be made subsidiary to almost any other business, and even a recreation as well as a source of profit.

In this book the endeavor has been, even at the risk of repetition, to make the best methods as plain as possible. The facts herein presented are the results of my own practical experience and observation, together with those obtained by extensive reading, travel and correspondence.

To Mr. Charles A. Dana, the proprietor of the Dosoris mushroom cellars and estate, I am greatly indebted for opportunities to prepare this book. For the past eight years everything has been unstintedly placed at my disposal by him to grow mushrooms in every way I wished, and to experiment to my heart's content.

To Mr. William Robinson, editor of *The Garden*, London, I am especially indebted for many courtesies— permission to quote from *The Garden*, "Parks and Gardens of Paris," and his other works, and to illustrate the chapters in this book on Mushroom-growing in the London market gardens and the Paris caves, with the original beautiful plates from his own books.

The recipes given in the chapter on Cooking Mushrooms, except those prepared for this work by Mrs. Ammersley, although based on the ones given by Mr. Robinson, have been considerably modified by me and repeatedly used in my own family.

My thanks are also due to Mr. John F. Barter, of London, the largest grower of mushrooms in England, for information given me regarding his system of cultivation; to Mr. John G. Gardner, of Jobstown, N. J., one of the most noted growers for market in this country, for facilities allowed me to examine his method of raising mushrooms; and to Messrs. A. H. Withington, Samuel Henshaw, George Grant, John Cullen, and other successful growers for assistance kindly rendered.

WILLIAM FALCONER.

DOSORIS, L. I., 1891.

TABLE OF CONTENTS.

TABLE OF CONTENTS.

TABLE OF CONTENTS.

ILLUSTRATIONS

Mushrooms, How to Grow Them.

CHAPTER I.

Market Gardeners.—The mushroom is a highly prized article of food which can be as easily grown as many other vegetable products of the soil—and with as much pleasure and profit. Below it is shown, in particular, that this peculiar plant is singularly well adapted to the conditions that surround many classes of persons, and by whom the mushroom might become a standard crop for home use, the city market, or both. It is directly in their line of business; is a winter crop, requiring their care when outdoor operations are at a standstill, and they can most conveniently attend to growing mushrooms. They have the manure needed for their other crops, and they may well use it first for a mushroom crop. After having borne a crop of mushrooms it is thoroughly rotted and in good condition for early spring crops; and for seed beds of tomatoes, lettuces, cabbages, cauliflowers, and other vegetables, it is the best kind of manure.

Years ago market gardening near New York in winter was carried on in rather a desultory way, and the supply of salads and other forced vegetables was limited and mostly raised in hotbeds and other frames, and prices

9

ran high. But of recent years our markets in winter have been so liberally supplied from the Southern States, that, in order to save themselves, our market gardeners have been compelled to take up a fresh line in their business, and renounce the winter frames in favor of greenhouses, and grow crops which many of them did not handle before. These greenhouses are mostly long, wide (eighteen to twenty feet), low, hip-roofed (30°) structures. In most of them the salad beds are made upon the floor, and the pathways are sunken a little so as to give headroom in walking and working. Others of these greenhouses are built a little higher, and middle and side benches are erected within them, as in the case of florists' greenhouses, and with the view of growing salad plants on these benches as florists do carnations, and mushrooms under the benches. The mushrooms are protected from sunlight by a covering of light boards, or hay, or the space under the benches is entirely shut in, cupboard fashion, with wooden shutters. The temperature is very favorable for mushrooms,—steady and moderately cool, and easily corrected by the covering-in of the beds ; and the moisture of the atmosphere of a lettuce house is about right for mushrooms. In such a house the day temperature may run up, with sunshine, to 65° or 70° in winter, but an artificial night temperature of only 45° to 50° is maintained. Under these conditions, with the beds about fifteen inches thick, they should continue to yield a good crop of short-stemmed, stout mushrooms for two or three months, possibly longer.

Besides growing the mushrooms in greenhouses our market gardeners are very much in earnest in cultivating them in cellars. Some of these cellars are ordinary barn cellars. others—large and commodious—have been built under barns and greenhouses, purposely for the cultivation of mushrooms. Several of these mushroom cellars may be found on Long Island between Jamaica and Woodhaven.

Florists.—In midwinter the cut flower season is at its height and the florist endeavors to make all the money out of his greenhouses that he possibly can ; every available inch of space exposed to the light is occupied by growing plants, and under the benches alongside of the pathways dahlias, cannas, caladiums, and other tubers and bulbs are stored, also ivies, palms, succulents and the like. In order that the plants may be more fully exposed to the sunlight, they are grown on benches raised above the ground so as to bring them near to the glass ; and the greenhouse seems to be full to overflowing. But right here we have the best kind of a mushroom house. The space under the benches, which is nearly useless for other purposes, is admirably adapted for mushroom beds, and the warmth and moisture of the greenhouse are exceptionally congenial conditions for the cultivation of mushrooms. Florists need the loam and manure anyway, and these are just as good for potting purposes—better for young stock—after having been used in the mushroom beds than they were before, so that the additional expense in connection with the crop is the labor in making the beds and the price of the spawn. Mushrooms are not a bulky crop ; they require no space or care in summer, are easily grown, handled, and marketed, and there is always a demand for them at a good price. If the crop turns out well it is nearly all profit ; if it is a complete failure very little is lost, and it must be a bad failure that will not yield enough to pay for its cost. Why should the florist confine himself to one crop at a time in the greenhouse when he may equally well have two crops in it at the same time, and both of them profitable ? He can have his roses on the benches and mushrooms under the benches, and neither interferes with the other. Let us take a very low estimate :   Iu a greeuhouse a hundred feet long make a five foot wide mushroom bed under the main bench ; this will give 500 square feet of bed, and half a pound to the foot will give 250 pounds of mushrooms, which, sold at fifty cents a pound net, brings $125.   This amount the florist would not have realized without growing the mushrooms.

Private Gardeners.—It is a part of their routine duty, and success in mushroom growing is as satisfactory to themselves as it is gratifying to their employers. Fresh mushrooms, like good fruit and handsome flowers, are a product of the garden that is always acceptable. One of the principal pleasures in having a large garden and keeping a gardener consists in being able to give to others a part of the choicest garden products.

In most pretentious gardens there is a regular mushroom house, and the growing of mushrooms is an easy matter ; in others there is no such convenience, and the gardener has to trust to his own ingenuity where and how he is to grow the mushrooms.   But so long as he has an abundance of fresh manure he can usually find a place in which to make the beds.   In the tool-shed, the potting-shed, the wood-shed, the stoke-hole, the fruit-room, the vegetable-cellar, or in some other out-building he can surely find a corner ; or, handier still, convenient room under the greenhouse benches, where he can make some beds.   Failing all of these he can start in August or September and make beds outside, as the London market gardeners do.

In fruit-forcing houses, especially early graperies, gardeners have a prejudice against growing any other plants than the grapevines lest red spiders, thrips, or mealy bugs are introduced with the plants, but in the case of mushrooms no such grounds are tenable.   As the vines have yielded their fruit by midsummer and ripened their wood early so as to be ready for starting into growth again in December or January, the grapery is kept cool and ventilated in the fall and early winter, but this need not interfere with the mushroom crop. Box up the beds or make them in frames inside the grapery ; the warm manure will afford the mushrooms heat enough until it is time to start the vines, when the increased temperature and moisture of the house will be in favor of the mushrooms because of the declining heat in the manure beds. The mushrooms have no deleterious effect whatever upon the vines, nor have the vines upon the mushrooms.

Village People and Suburban Residents.—Those who keep horses should, at least, grow mushrooms for their own family use and, if need be, for market as well. They are so easily raised, and they take up so little space that they commend themselves particularly to those who have only a village or suburban lot, and, in fact, only a barn. And they are not a crop for which we have to make a great preparation and need a large quantity of manure. No matter how small the bed may be, it will bear mushrooms ; and if we desire we can add to the bed week after week, as our store of manure increases, and in this way keep up a continuous succession of mushrooms. A bed may be made in the cow-house or horse-stable, the carriage-house, barn-cellar, wood-shed, or house-cellar ; or if we can not spare much room anywhere, make a bed in a big box and move it to where it will be least in the way. But the best place is, perhaps, the cellar. An empty stall in a horse-stable is a capital place, and not only affords room for a full bed on the floor, but for rack-beds as well.

Farmers.—No one can grow mushrooms better or more economically than the farmer. He has already the cellar-room, the fresh manure and the loam at home, and all he needs is some spawn with which to plant the beds. Nothing is lost. The manure, after having been used in mushroom beds, is not exhausted of its fertility, but, instead, is well rotted and in a better condition to apply to the land than it was before being prepared for the mushroom crop. The farmer will not feel the little labor that it takes. There is no secret whatever connected with it, and skilled labor is unnecessary to make it successful. The commonest farm hand can do the work, which consists of turning the manure once every day or two for about three weeks, then building it into a bed and spawning and molding it. Nearly all the labor for the next ten or twelve weeks consists in maintaining an even temperature and gathering and marketing the crop.

Many women are searching for remunerative and pleasant employment upon the farm, and what can be more interesting, pleasant and profitable work for them than mushroom-growing? After the farmer makes up the mushroom bed his wife or daughter can attend to its management, with scarcely any tax upon her time, and without interfering with her other domestic duties. And it is clean work; there is nothing menial about it. No lady in the land would hesitate to pick the mushrooms in the open fields, how much less, then, should she hesitate to gather the fresh mushrooms from the clean beds in her own clean cellar? Mushrooms are a winter crop; they come when we need them most. The supply of eggs in the winter season is limited enough, and pin-money often proportionately short; but with an insatiable market demand for mushrooms all winter long, at good prices, no farmer's wife need care whether the hens lay eggs at Christmas or not. When mushroom-growing is intelligently conducted there is more money in it than in hens, and with less trouble.

CHAPTER II.

Underground Cellars.—Mushrooms require a uniform moderately low temperature and moist atmosphere, and will not thrive where draughts, or sudden fluctuations of temperature or moisture prevail. Therefore an underground cellar is the best of all structures in which to grow mushrooms. The cellar is everybody's mushroom house.

Cellars are under dwellings, barns, and often under other out-buildings. These cellars are imperative for domestic purposes, for storing apples, potatoes and other root crops and perishable produce; and for these uses we need to make them frost proof and dry. These cellars are ideal mushroom houses, and any one who has a good cellar can grow mushrooms in it. In fact, our market gardeners who are making money out of mushrooms find it pays them to excavate and build cellars expressly for growing mushrooms. Indeed, some of our market gardeners who have never grown a mushroom or seen one grown, but who know well that some of their neighbors are making money out of this business, instinctively feel that the first step in mushroom-growing is a cellar. It is almost incredible how secretly the market growers guard everything in connection with mushroom-growing from the outside world, and even from one another; in fact, in some cases their next-door neighbors and life-long intimate friends have never been inside their mushroom cellars.

If a cellar is to be wholly devoted to mushroom-grow-

15

ing it should be made as warm as possible with double windows, and double doors, where the entrance is from the outside, but if from another building single doors will suffice. A chimney-like shaft or shafts rising from the ceiling should be used as ventilators in winter, when we can not ventilate from doors or windows; indeed, side ventilation at any time when the beds are in bearing condition is rather precarious. There should be some indoor way of getting into the cellar, as by a stairway from the building above it. Also an easy way of getting in fresh materials for the beds, and removing the exhausted material. This is, perhaps, best obtained by having a door that opens to the outside, or a moderately large one from the building above.

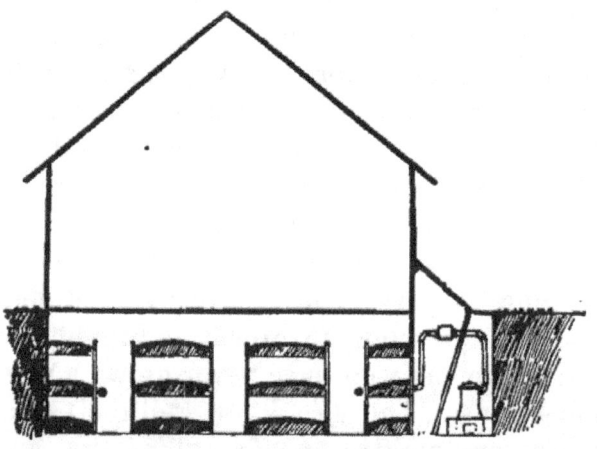

FIG. 1. MUSHROOM CELLAR UNDER A BARN.

The interior arrangement of the cellar is a matter of choice with the grower, but the simplest way is to have beds three or four feet wide around the inside of the walls, and beds six feet wide, with pathways two, or two and one-half feet wide between them running parallel along the middle of the cellar. Above these floor-beds, shelf-beds in tiers of one, two, or three, according to the height of the cellar, may be formed, always leaving a space of two and one-half or three feet between the bottom of one bed and the bottom of the next. This is very necessary, in order to admit of making and tending the beds and gathering the crop, and emptying the beds when they are exhausted.

Provision should also be made for the artificial heating of these cellars, and room given for the heating pipes wherever they are to run. But wherever fire heat is used in heating these cellars, if practicable, the furnace itself should be boxed off, by a thin brick wall, from the main cellar, and the pipes only introduced. This does away with the dust and noxious gas, and modifies the parching heat.

But in a snug, warm cellar, artificial heat is not absolutely necessary. We can grow capital crops of mushrooms in such a cellar without any furnace heat, simply by using a larger body of material in making the beds,— enough to maintain a steady warmth for a long time. But this, observe, is a waste of material, for no more mushrooms can be grown in a bed two feet thick than in one a foot thick. In an unheated cellar the mushrooms grow large and solid, but they do not come so quickly nor in such large numbers as in a heated one. And a little artificial warmth has the effect of dispelling that cold, raw, damp air peculiar to a pent-up cellar in winter, and purifies the atmosphere by assisting ventilation.

Instead of using box beds, some growers spread the bed all over the floor of the cellar, and leave no pathway whatever, stepping-boards or raised pathways being used instead. Of course, in these instances, no shelf beds are used. Others make ridge beds all over the cellar floor, as the Parisians do in the caves. The ridges are two feet wide at bottom, two feet high, and six or

2

eight inches wide at top, and there is a foot alley between
them.   Here, again, no shelf beds are used.

One of the chief troubles with flat-roofed mushroom
cellars is the drip from the condensed moisture rising
from the beds, and this is more apparent in unheated
than in heated cellars,—the wet gathers upon the ceiling and, having no slope to run off, drips down again.
Oiled paper or calico strung along $\wedge$ wise above the
upper beds protects them perfectly ; whatever falls upon
the passage-ways upon the floor does no harm.

In any other outhouse cellar, as well as in one completely given over to this use, we can make up beds and
grow good mushrooms.   Mr. James Vick told me that at
his seed farm near Rochester he raises many mushrooms
in winter in his potato cellars ; and so can any one in
similar places.   Mr. John Cullen, of South Bethlehem,
Pa., a very successful cultivator, tells me that his present
mushroom cellar used to be a large underground cistern,
but with a little fixing, and opening a passage-way to it
from a neighboring cellar, he has converted it into an
excellent cellar for mushrooms, and surely the immense
crops that I have seen in that cave of total darkness justify his good opinion of it.

In Dwelling House.—The cellar of a dwelling
house is a capital place for mushroom beds, and can be
used in whole or part for this purpose.   In the case of
private families who wish to grow a few mushrooms only
for their own use it is not necessary to devote a whole
cellar to it ; but partition off a part of it with boards
and make the beds in this.   Or make a bed alongside of
the wall anywhere and box it in to protect it from cold
and draughts, and mice and rats.   You can have shelves
above it for domestic purposes, just as you would in any
other part of the cellar.   Bear in mind that mushrooms
thrive best in an atmospheric temperature of from 50° to
60°, and if you can give them this in your house-cellar

you ought to get plenty of good mushrooms. But if such a high temperature can not be maintained without impairing the usefulness of the cellar for other purposes, box up the beds tightly, and from the heat of the bed itself, when thus confined, there usually will be warmth enough for the mushrooms, but if not spread a piece of old carpet or matting over the boxing.

The beds may be made upon the floor, and flat, or ridged, or banked against the wall, ten or twelve inches deep in a warm cellar, and fifteen to twenty inches or more deep in a cool cellar, and about three feet wide and any length to suit.

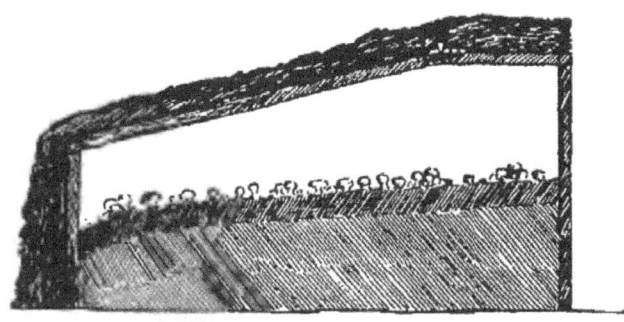

FIG. 2. BOXED-UP FRAME WITH STRAW COVERING.

The boxing may consist of any kind of boards for sides and ends, and be built about six or ten inches higher than the top of the beds, so as to give the mushrooms plenty headroom; the top of the boxing may be a lid hung on hinges or straps, or otherwise arranged, to admit of being easily raised or removed at will, and made of light lumber, say one-half inch thick boards. In this way, by opening the lid, the mushrooms are under observation and can be gathered without any trouble. When the lid is shut they are secure from cold and vermin. Thus protected the cellars can be ventilated without interfering with the welfare of the mushrooms. A light wooden frame covered with calico or oiled paper would also make a good top for the boxing, only it would not be proof against much cold, or rats or mice. If desirable, in warm cellars, shelf beds could be built above the floor beds, but in cool, airy cellars this would not be advisable.

Manure beds in the dwelling-house cellar may seem highly improper to many people, but in truth, when rightly handled, these beds emit no bad odor. The manure should be prepared away from the house, and when ready for making into beds it can be spread out thin, so as to become perfectly cool and free from steam. When it has lain for two days in this condition it may be brought into the cellar and made into beds. Having been well sweetened by previous preparation, it is now cool and free from steam, and almost odorless; after a few days it will warm up a little, and may then be spawned and earthed over at once. Do not bury the spawn in the manure, merely set it in the surface of the manure; this saves the spawn from being destroyed by too great a heat, should the bed become unduly warm. This, if the manure has been well prepared, is not likely to occur. The coating of loam prevents the escape of any further steam or odor from the manure.

On the 14th of January last, Mr. W. Robinson, editor of the London *Garden*, in writing to me, mentioned the following very interesting case of growing mushrooms in the cellar of a dwelling house: "I went out the other day to see Mr. Horace Cox, the manager of the *Field* newspaper, who lives at Harrow, near the famous school. His house is heated by a hot-water system called Keith's, and the boiler is in a chamber in the house in the basement. The system interested me and I went down to see the boiler, which is a very simple one worked with coke refuse. However, I was pleased to see all the floor of the room not occupied by the boiler covered with little flat mushroom beds and bearing a very good crop. Truth to tell, I used to fear growing mushrooms in dwelling houses might be objectionable in various ways; but this instance is very interesting, as there is not even the slightest unpleasant smell in the chamber itself. The beds are small, scarcely a foot high, and perfectly odorless; so that it is quite clear that one may cultivate mushrooms in one's house, in such a case as this, without the slightest offence."

Mr. Gardner's Method.—Mr. J. G. Gardner, of Jobstown, N. J., uses an ordinary cellar, such as any farmer in the country has, and the little that has been done to it to darken the windows and make them tight, so as to render them better for mushrooms, any farmer with a hand-saw, an ax, a hammer and a few nails and some boards can do. Mr. Gardner is a market gardener, and has not the amount of fresh manure upon his own place that he needs for mushroom-growing, but he buys it, common horse manure, in New York, and it is shipped to him, over seventy miles, by rail. And this pays; and if it will pay a man to get manure at such a cost for mushroom-growing, how much more will mushroom-growing pay the farmer who has the cellar and the manure as well? Mr. Gardner raises mushrooms, and lots of them. When I visited him last November, instead of trying to hide anything in their cultivation from me, he took particular pains to show and explain to me everything about his way of growing them. And he assures me that by adopting simple means of preparing the manure and "fixing" for the crop, and avoiding all complicated methods, one can get good crops and make fair profits.

His cellar is sixty feet long, twenty-four feet wide, and nine feet high from floor to ceiling. The floor is an earthen one, but perfectly dry. It is well supplied with window ventilators and doors, and in the ceiling in the middle of the cellar opens a tall shaft or chimney-like ventilator that passes straight up through the roof above. While the beds are being made full ventilation by doors, windows and shaft is given, but as soon as there is any sign of the mushrooms appearing all ventilators except the shaft in the middle are shut and kept closed.

The bed occupies the whole surface of the cellar floor and was all made up in one day. As a pathway, a single row of boards is laid on the top of the bed, running lengthwise along the middle of the cellar from the door to the farther end, and here and there between this narrow path and the walls on either side a few pieces of slate are laid down on the bed to step upon when gathering the mushrooms. Here is the oddest thing about Mr. Gardner's mushroom-growing. He does not give the manure any preparatory treatment for the beds. He hauls it from the cars to the cellar, at once spreads it upon the floor and packs it solid into a bed. For example, on one occasion the manure arrived at Jobstown, July 8th ; it was hauled home and the bed made up the same day, and the first mushrooms were gathered from this bed the second week in September,—just two months from the time the manure left the New York or Jersey City stables. The bed was fifteen inches thick. In making it the manure was first shaken up loosely to admit of its being more evenly spread than if pitched out in heavy forkfuls, and it was then tramped down firmly with the feet. The bed was then marked off into halves. On one half (No. 1) a layer of a little over three inches of loam was at once placed over the manure ; on the other half (No. 2) no loam was used at this time, but the manure on the surface of the bed—about three inches deep—was forked over loosely. Twelve days after having been put in the temperature of the bed No. 2, three inches deep, was 90°, and then it was spawned.

On the next day the soil from bed No. 1, spawned four days earlier, was thrown upon bed No. 2, and then part of the soil that was thrown on No. 1 was thrown back again on No. 2, so that now a coating of loam an inch and a half deep covered the whole surface of the bed. When finished the surface was tamped gently with a tamper with a face of pine plank sixteen inches long by twelve inches wide. Mr. Gardner does not believe in the alleged advantages of a hard-packed surface on the mushroom bed, but is inclined to favor a moderately firm one.

He uses the English brick spawn, which is sold by our seedsmen. He has tried making his own spawn, but owing to not having proper means for drying it, he has had rather indifferent success.

Almost all growers insert the pieces of spawn about two to three inches under the surface of the manure, one piece at a time, and at regular intervals of nine inches or thereabouts apart each way—lengthwise and crosswise. But here, again, Mr. Gardner displays his individuality. He breaks up the spawn in the usual way, in pieces one or two inches square. Of course, in breaking it up there is a good deal of fine particles besides the lumps. With an angular-pointed hoe he draws drills eighteen inches apart and two and one-half to three inches deep lengthwise along the bed, and in the rows he sows the spawn, as if he were sowing peach stones, or walnuts, or snap beans, and covers it in as if it were seeds.

Mr. Gardner regards 57° as the most suitable temperature for a mushroom house or cellar, and, if possible, maintains that without the aid of fire-heat. He has hot-water pipes connected with the contiguous greenhouse heating arrangement in his cellar, but he never uses them for heating the mushroom cellar except when obliged to. By mulching his bed with straw he gets

along without any fire-heat, but this is very awkward
when gathering the mushrooms.

After the bed has borne a little while it is top-dressed
all over with a half-inch layer of fine soil. Before using,
this soil has been kept in a close place—pit, frame, shed,
or large box—in which there was, at the same time, a
lot of steaming-hot manure, so that it might become
thoroughly charged with mushroom food absorbed from
the steam from the fermenting material.

Should any portion of the bed get very dry, water of a
temperature of 90° is given gently and somewhat sparingly through a fine-spraying waterpot rose, or syringe.
Enough water is never given at any one time to penetrate through the casing into the manure below or the
spawn in the manure. But rather than make a practice
of watering the beds, Mr. Gardner finds it is better to
maintain a moist atmosphere, and thus lessen the necessity for watering.

Mr. Gardner firmly believes that the mushrooms de-
rive much nourishment from the "steam" of fermenting
fresh horse manure, and by using this "steam" in our
mushroom houses we can maintain an atmosphere almost
moist enough to be able to dispense with the use of the
syringe, and the mushrooms are fatter and heavier for
it. And he practices what he preaches. In one end of
his mushroom cellar he has a very large, deep, open box,
half filled with steaming fresh horse-droppings, and once
or twice a day he tosses these over with a dung-fork, in
order to raise a "steam," which it certainly does. It is
also for this purpose that he introduces the loam so soon
when making the beds, so that it may become charged
with food that otherwise would be dissipated in the
atmosphere.

There is a marked difference between the mushrooms
raised from the French flake spawn and those from the
English brick spawn, but he has never observed any dis-

tinct varieties from the same kind of spawn. Sometimes a few mushrooms will appear that are somewhat differently formed from those of the general crop, but this he regards as the result of cultural conditions rather than of true varietal differences.

His last year's bed began bearing early in November, and continued to bear a good crop until the first of May. After that time, no matter what the crop may be, the mushrooms become so infested with maggots as to be perfectly worthless, and they are cleared out. It is on account of the large body of manure in the bed, and the low, genial, and equable temperature of the cellar that the beds in this house always continue so long in good cropping condition.

Some years ago the mushrooms were not gathered till their heads had opened out flat, but nowadays the marketmen like to get them when they are quite young and before the skin of the frill between the cup and the stem has broken apart. A good market is found in New York, Philadelphia and Boston.

Mr. Denton's Method.—Mr. W. H. Denton, of Woodhaven, L. I., is an extensive market gardener about ten miles from New York. During the summer months he grows outdoor vegetables for the New York and Brooklyn markets, and in winter mushrooms in cellars. He has no greenhouses. Under his barns he has two large cellars which he devotes entirely to mushroom-growing in winter. The cellars are seven and one-half feet high inside; the beds five feet wide, nine inches deep, two feet apart, and run parallel to one another the whole length of the cellar. The beds are three deep, that is, one bed is made upon the floor, and the other two, rack or shelf fashion, are made above the floor bed, and two and one-half feet apart from the bottom of the one bed to the bottom of the one above it. The shelves altogether are temporary structures built of ordinary rough

scantling and hemlock boards, and the beds are all one
board deep.

A common iron stove and string of sheet iron smoke
pipes are used for heating the cellars. But he tells me
the parching effect is very visible on the beds, it dries
them up on the surface very much, and he has to
sprinkle them frequently with water to keep them moist
enough. During the late summer and fall months, on
his return trips from the Brooklyn markets, Mr. Denton
hauls home fresh horse manure from the City stables.
All that he can put on a wagon costs him about twenty-
five cents; and this is what he uses for mushrooms. He
prepares it in a large open shed just above the cellar,
and when it is fit for use he adds about one-third of its
bulk of loam. The loam is the ordinary field soil from
his market garden. He tells me he has better success
with beds made up in this way than when manure alone
is used. We all know how very heavily market gardeners manure their land, also how vigorously most writers
on mushroom culture denounce the use of manure-fatted
loam in mushroom beds, but here is Mr. Denton, the
most successful grower of mushrooms for market in the
neighborhood of New York, practicing the very thing
that is denounced! While he likes good lively manure
to begin with he is very careful not to use it soon enough
to run any risk of overheating in the beds. The loam in
the manure counteracts this strong heating tendency,
also with the loam mixture the shelf-beds can be built
much more firmly than with plain manure on the springy
boards. When the temperature falls to 90° he spawns
the beds.

He uses both French and brick spawn, but leans with
most favor to the latter, of which in the fall 1889 he used
400 lbs. He markets from 1700 to 2500 lbs. of mushrooms a year from these two cellars. Mr. Denton believes emphatically in cleanliness in the mushroom cel-

lar, and ascribes his best successes to his most thorough cleaning. Every summer he cleans out his cellars and limewashes all over.

Mr. Van Siclen's Method.—Mr. Abram Van Siclen, of Jamaica, L. I., also grows mushrooms very extensively in underground cellars, whose arrangements do not differ materially from those of Mr. Denton's, except in his manner of heating. He runs an immense greenhouse vegetable-growing establishment, as well as a summer truck farm, and uses hot water heating apparatus, also smoke flues as employed ordinarily in greenhouses, especially lettuce houses. The sheet iron pipes, except in squash houses, he does not hold in much favor.

The Dosoris Mushroom Cellar.—This is a subterranean tunnel or cellar that was excavated and arched some ten years ago, expressly for the cultivation of mush-

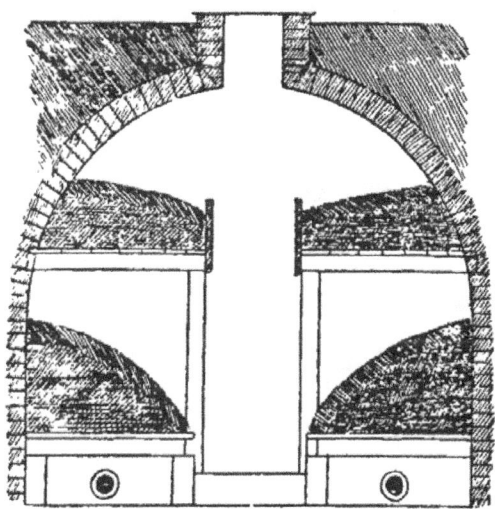

FIG. 3. CROSS-SECTION OF THE DOSORIS MUSHROOM CELLAR.

rooms. It is situated in an open, sunny part of the garden, and its extreme length from outside of end walls is eighty-three feet: but of this space nine feet at either

end are given up to entrance pits and a heating apparatus; and the full length of the mushroom cellar proper
inside the inner walls is sixty-three feet.   The walls and
arch are of brick, and the top of the arch is two and
one-half feet below the surface of the soil.   This tunnel
or arch is seven feet high in the middle and eight feet
wide within, but a raised two-feet-wide pathway along
the middle lessens the height to six and one-half feet.
Between this pathway and the sides of the building there
is only an earthen floor, but it is quite dry, as the cellar
is perfectly drained.   Three ventilators sixteen feet
apart had been built in the top of the arch, but this was
a mistake, as the condensation in the cellar in winter
from these ventilators always keeps the place under them
cold and wet and rather unproductive.   One tall wooden
chimney-like shaft would have been a better ventilator
than the three ventilating holes now there, which are
covered over with an iron and glass grating.

At one end of the house and behind the stairs descend-
ing into the pit is the heating apparatus, from which a
four-inch hot-water pipe passes around inside the house

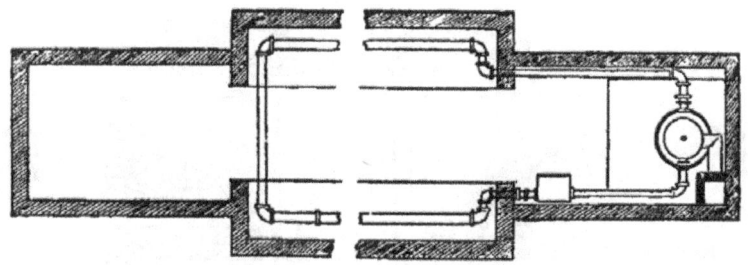

FIG. 4.   GROUND PLAN OF THE DOSORIS CELLAR.

near the wall and only four inches above ground.   A
three-feet wide hemlock flooring for the bed to rest on
is laid along each side and about four inches above the
pipe, leaving the aperture between the earth floor and
the bottom of the bed along the pathway open for the

escape of the artificial heat. One might think that the hot water pipe under, and so near the bed, would dry it up and destroy it, but such is not the case. In a cellar of this kind very little fire heat is needed to maintain the required temperature, and I do not know where else the pipes could be put where they would do the work any better and be more out of the way.

These beds, for convenience in building them, spawning them, molding them over, gathering the crop and watering the beds, and removing the manure after the beds are exhausted, are built against the wall and with a rounded face, thus giving a three and one-half feet wide surface of bed in place of one three feet wide, were it built flat. This gain in superficial area is not so important as it might seem, for the part immediately next to the edge of the pathway seldom yields very much. Above these beds a string of shelf beds is arranged which runs the full length of both sides of the cellar. From the floor of the under bed to the floor of the top bed is three feet, and the upper beds are just as wide as the lower ones. The shelves for the beds are temporary affairs, put up and taken down every year. The crossbars rest in sockets in the wall made by cutting out half a brick every four feet along the wall, and on upright strips or feet one and one-fourth by four inches wide, or two by three inches, set under the inside ends of the cross-bars and resting on the cement floor close up against the lower bed. By having this foot end a quarter of an inch higher than the wall end the heavy weight of the bed is thrown toward the wall. Loose hemlock boards set close together form the flooring, for there is no need of nailing any of them except the one next to the upright face board, which is ten inches wide, and nailed along the front, by the pathway, to the posts and shelf board. By tilting the weight to the wall the upright board is firm enough to hold its place against any

pressing out in building the beds. The supporting legs
of the shelves are also nailed to the face board of the
lower bed, and this holds them perfectly solid in place.
The shelf beds are eight inches deep at front, but can
be made of any depth desired against the walls at the
back. The cold wall has no injurious effect upon the
bearing of the bed, and many fine mushrooms grow close
against the walls.

The entrance pits are nine and one-half feet deep from
ground level, three feet eight inches wide, nine feet
long, and are covered over with folding doors on strong
hinges, and descended into by means of wooden movable stairs. These dimensions are needed at the end
where the heating apparatus is placed, but at the other
end, although it is convenient in handling the manure, a
space two or three feet less would have answered just as
well. A close door at either end of the mushroom cellar
proper separates it from the end pits. The cellar is
divided in the middle by a partition. This gives, when
it is in full working order, eight beds, each thirty-one
and one-half feet long, or a continuous run of 252 feet
or 756 square feet of surface, and as the beds are renewed twice a year this gives 504 running feet of bed,
or 1512 square feet of surface. A common average crop
is three-fifths of a pound of mushrooms to the square
foot of bed, and a good fair average is four-fifths of a
pound. This would give over a thousand pounds of
mushrooms a season from this cellar when it is in full
running capacity. But as the aim is to have a steady
supply of mushrooms from October until May, and not a
flush at any one time and a scarcity at another, only two
beds are made at a time, allowing a month to intervene
between every two.

For the two beds, No. 1, preparing the manure begins
in July, the beds are made up in August, and gathering
of the crop commences in October; work on the two

beds, No. 2, begins in August, the beds are made up in September, and the mushrooms gathered in November ; preparing for the two beds, No. 3, begins in September, the beds are made up in October, gathering commences in December ; for the two beds, No. 4, work begins in October, the beds are made up in November, and the crop is gathered in January ; for the two beds, No. 5 (No. 1 renewed), work begins in November, the beds are made up in December, and the crop is gathered in February ; for the two beds, No. 6 (No. 2 renewed), work begins in December, the beds are made up in January, and the crop is gathered in March ; for the two beds, No. 7 (No. 3 renewed), work begins in January, the beds are made up in February, and the crop is gathered in April ; for the two beds, No. 8 (No. 4 renewed), work begins in February, the beds are made up in March, and the mushrooms gathered in May. After this time of year the summer heat renders mushroom-growing uncertain, and the maggots destroy the mushrooms. This system allows each bed a bearing period of two months. After yielding a crop for some seven to nine weeks the beds are pretty well exhausted and hardly worth retaining longer. They might drag along in a desultory way for weeks, but as soon as they stop yielding a paying crop we clear them out and start afresh.

And when the mushroom season is closed we lift out and remove the manure, clean the boards used in shelving, and give the cellar a thorough cleaning,—whitewash its walls and paint its woodwork with kerosene to destroy noxious insects and fungi.

The heating apparatus consists of one of Hitchings' base-burner boilers with a four-inch hot-water pipe that passes around inside the cellar, and it deserves special mention because of its economy, efficiency, and the satisfaction it gives generally. This boiler needs no deep or

spacious stokehole.   Here it is set under the stairway in
a pit four and one-half feet long, by three feet wide, by
eighteen inches deep; it is not in the way, and there is
plenty of room to attend to it.   The heater, like a common parlor stove, has a magazine for the supply of coal.

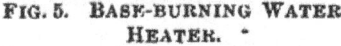

FIG. 5.   BASE-BURNING WATER   FIG. 6.   VERTICAL SECTION.
        HEATER. ·

It has a double casing with the water space between and
down to the bottom of it, so that when set in a shallow
pit there is no difficulty whatever about the circulation
of the water in the pipes.   The hot water passes from
the boiler to an open iron tank placed two feet above it,
as shown in the engraving, and thence down through a
perpendicular pipe till it reaches and enters the horizontal pipes that pass around the cellar and, returning,
enters the boiler again near its base.   The boiler and
pipes are filled from this tank, which should always be
kept at least half full of water, and looked into every
day when in use, so that when the water gets lower than
half full it may be filled up again: About 134 running
feet of four-inch pipe are included inside the cellar
(sixty-four feet on each side and six feet across at further
end); this gives 134 square feet of heating surface, or a

proportion of about a square foot of heating surface for every fifteen cubic feet of air space in the cellar. This proportion is more than ample in the coldest weather, but beneficial in so far that there is no need to fire hard to maintain the proper temperature. A three-inch pipe would have given heat enough, but the heat would not have been so steady. Both nut and stove coal is used in this heater, and in the severest winter weather it burns not more than a common hodful in twenty-four hours. It is so easily regulated that the temperature of the cellar day or night, or in mild or severe weather, never varies more than three degrees, namely from 57° to 60°.

In a close underground cellar where the temperature in midwinter without any artificial heat does not fall below 40° or 45° it is an easy matter, with such a heater as this is, to maintain any desired temperature. If the grates are renewed now and then, the heater should last in good condition for twenty years. With the ordinary stove there is danger of fire, of escaping gas and of sudden changes of temperature, and the evil influence of a dry, parching heat—just what mushrooms most dislike —is ever present. The first cost of a hot water apparatus may be more than that of an old stove and sheet iron pipes, but where mushrooms are grown extensively, as a matter of economy, efficiency, and convenience, the advantages are altogether on the side of the hot water apparatus. Furthermore, hot water pipes can be run where it would be unsafe to put smoke pipes.

CHAPTER III.

GROWING MUSHROOMS IN MUSHROOM HOUSES.

A mushroom house is a building erected purposely for mushroom culture. It may be wholly or partly above ground, and built of wood, brick, or stone, and extend to any desired dimensions. But a few general principles should be borne in mind. Mushrooms in houses are a winter and not a summer crop, and they are impatient of sudden changes of temperature and of a hot or arid atmosphere. Therefore, build the houses where they will be warm and well-sheltered in winter, so as to get

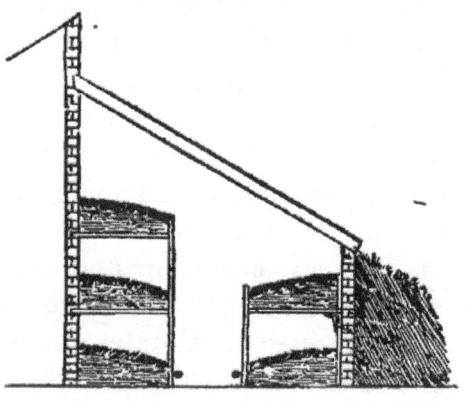

FIG. 7. MUSHROOM HOUSE BUILT AGAINST A NORTH-FACING WALL.

the advantage of the natural warmth, and spare the artificial heat. They should be entered from an adjoining building, or through a porch on the south side, so as to guard against cold draughts or blasts in winter when the door would be opened in going into or coming out

34

of the house. At the same time, do not lose sight of convenience in handling the manure, either in bringing it into the house or taking it out, and with this in view it may be necessary to have a door opening to the outside. All outside doors should be double and securely packed around in winter. Side window ventilators are not necessary, at the same time they are useful in the early part of the season and in summer time; they should be double and tightly packed in winter. The walls, if made of brick, should be hollow, if of wood, double; indeed, walls built as if for an ice house are the very best for a mushroom house, and should be banked with earth, tree leaves, or strawy manure in winter, to help keep the interior of the house a little warmer.

The floor should be perfectly dry; that is, so well drained that water will not stand upon it, but it is

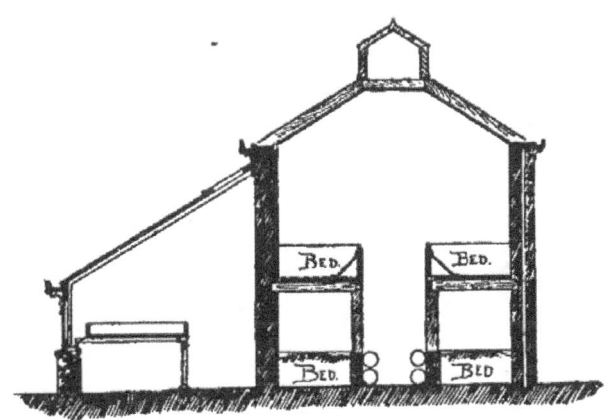

FIG. 8. SECTION OF MRS. C. J. OSBORNE'S MUSHROOM HOUSE.

immaterial whether the floor is an ordinary earthen one or of wood or cement.

The roof should be double and always sloping,—never flat. The hoar frost that appears in severe weather inside a single roof is likely to melt as the heat of the day increases, and this cold drip falling upon the beds below is very prejudicial to the mushroom crop. A double roof saves the beds from this drip, and it also renders the house warmer, and less fire is needed to maintain the requisite temperature. One might think that a single roof like that of a dwelling house, and then a flat ceiling under it, would be equivalent to a double sloping roof, but it is not. The moisture arising from the interior of the house condenses upon the flat ceiling, and the water, having no way of running off, drips down upon the beds. With a sloping ceiling or inside roof the water runs down the ceiling to the walls. A very pointed example of this may be seen in Mrs. C. J. Osborne's

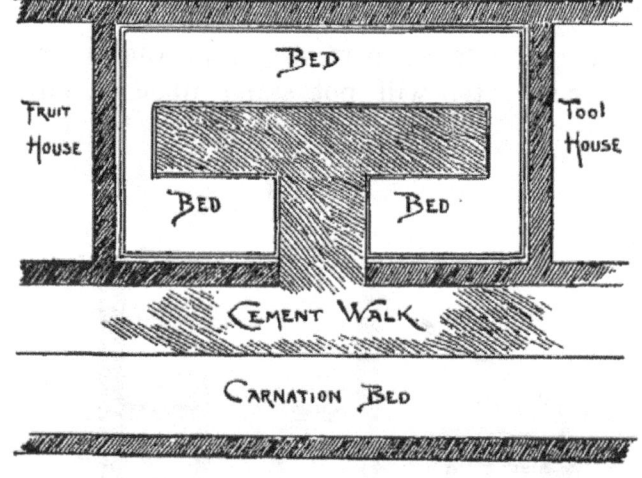

FIG. 9. GROUND PLAN OF MRS. OSBORNE'S MUSHROOM HOUSE.

excellent mushroom house at Mamaroneck, N. Y. It had been built in the most substantial manner, with a sloping roof and a flat ceiling under the roof, but so much annoyance was caused by the drip falling from it upon the beds below that her gardener had the flat ceiling removed and a sloping one built instead, and now it works splendidly, and a few months ago I saw as fine a crop of mushrooms in that house as one could wish to look at.

The interior arrangement of the mushroom house may resemble that of the mushroom cellar. Beds may be made alongside of the walls and, if there is room, also along the middle of the house, and shelves erected in the same way as in the cellar. But in the case of cold, thin outside walls, the shelf-beds should not be built close against them, but instead boxed off about two inches from the walls, so as to remove the beds from the chilling touch of the wall in winter. Economy may suggest the advisability of high mushroom houses, so that one may be able to build one shelf above another, until the shelves are two, three, or four deep. But this is a mistake. The artificial heat required to maintain a temperature of 55° in midwinter in a house built high above ground would be too parching and unsteady for the good of the mushrooms; besides, a second shelf is inconvenient enough, and when it comes to a third or a fourth the inconvenience would be too great, and over-reach any advantage hoped for in economy of space. An unheated mushroom house must be regarded as a shed, and treated similarly, as described in the following chapter.

In large, well appointed, private gardens, a mushroom house is considered an almost indispensable adjunct to the glasshouse establishment, and is generally built against the north-facing wall of a greenhouse. In this way it gets the benefit of the warm wall, and may be easily heated by introducing one or two hot-water pipes from the greenhouse system; besides, in winter the house may be entered from the glass house or adjacent shed, and in this way be exempted from the inclement breath of the frosty air that would be admitted in opening the outside door.

Mr. Samuel Henshaw's Mushroom House.—Mr.

Henshaw has raised mushrooms several years at his place on Staten Island. His mushroom house is nine feet wide and sixty feet long. One side is a brick wall and the other is double boarded.

The roof is of tin, in which there are three sashes each two by five feet, supplying ample light. At each end is a door giving convenient access to the interior, for carrying in and removing material without disturbing the bearing beds. In winter the roof is covered with a coating of salt hay, to pre-

FIG. 10.   INTERIOR VIEW OF MR. S. HENSHAW'S MUSHROOM HOUSE.

serve an equable temperature and prevent the moisture from condensing on the ceiling and falling in drops on the beds. The floor is of earth, which, when well drained, he thinks preferable to either brick or lumber. The floor is entirely covered with beds, no shelves or walks being used. This makes it necessary to step on the beds, but as no covering is employed it is always easy to avoid stepping on the clusters of young mushrooms, and so long as they are left uninjured the bed is seldom, if ever, impaired by the compacting effect of the treading. In order to maintain a necessary winter temperature of 60° a four-inch hot-water pipe extends the whole length of the house about two feet from the floor. On the other side of the brick wall is a greenhouse which, by keeping the wall warm, helps to keep the mushroom house warm. Mr. Henshaw divides this house into three equal beds. The part at the further end of the house is made up in the fall and comes into bearing in December; the middle part a month later to come in a month later, and the near end still a month later, to follow as another succession. Then, if need be, and he wishes to renew the bed at the further end of the house, he clears it out and supplies fresh material for the new bed.

CHAPTER IV.

GROWING MUSHROOMS IN SHEDS.

Any one who has a snug, warm shed, may have a good mushroom house, but it is imperative that the floor should be dry, and the roof water-tight. Of course a close shed, as a tool-house or a carriage-house, is better than an open shed, but even a shed that is open on the south side, if closely walled on the other sides, can also be made of good use for mushroom beds. While open sheds are good enough for beds that yield their crop before Christmas, they are ill-adapted for midwinter beds. The temperature of the interior of a mushroom bed should be about 60° during the bearing period, and the temperature of the surface of the bed 45° to 50° at least; if lower than that the mycelium has a tendency to rest, and the crop stagnates. Now this temperature can not be maintained in an open shed, in hard frosty weather, without more trouble than the crop is worth. The beds would have to be boxed up and mulched very heavily. And even in a close, warm shed, protection in this way would have to be given, but the bed should not be under the penetrating influence of piercing winds and draughts. The mushroom beds should therefore be made in the warmest parts of the warmest sheds.

The beds should be made upon the floor and as much to one side as possible, so as to be out of the way, and in form flat on the ground, or rounded up against the sides of the shed; in the latter case the house should be well

banked around on the outside with litter or tree leaves
or earth, so as to exclude frost from the lower part of
the walls, and thereby prevent the manure in the beds
from getting badly chilled. The beds should be made
deeper in a cool shed than in a cellar or warm mushroom house, so that they may retain their heat for a
long time.

Shelf beds should not be used in unheated sheds, because of the difficulty in keeping them warm in winter.
As a rule, shelf beds are not made as deep as are those
upon the floor; hence they do not hold their heat so
long. When cold weather sets in it is easy to box up
and cover over the lower beds to keep them warm, but
in the case of shelf beds, that are exposed above and
below, it is more trouble to protect them sufficiently
against cold than they are worth.

Generally speaking, the term shed is applied to unheated, simple wooden structures; for instance, the
wood-shed, the tool-shed, a carriage-house, or a hay-
barn. But we often use the name shed to designate
heated buildings, as the potting and packing sheds of
florists. Were it not that these heated sheds are simply
workrooms, and where there is a great deal of going out
and in, and, consequently, draughts and sudden and
frequent fluctuations of temperature, the treatment of
mushroom beds made in them would be the same as
that advised for regular mushroom houses; but as the
circumstances are somewhat different the treatment,
too, should not be the same. A warm potting shed is an
excellent place for mushroom beds. Here they should
be made under the benches and covered up in front with
thick calico, plant-protecting cloth, or light wooden
shutters, to exclude cold currents and sudden atmospheric changes, and guard against the beds drying too
quickly.

CHAPTER V.

Any one who has a greenhouse can grow mushrooms in it. And it does not matter what kind of greenhouse it is, whether a fruit house, a flower house, or a vegetable house, it is available for mushrooms. One of the advantages of raising mushrooms in a greenhouse is that they grow to perfection in parts of the greenhouse that are nearly worthless for other purposes; for instance, under the stages, where nothing else grows well, although rhubarb and asparagus might be forced there, and a little chicory and dandelion blanched.

Cool greenhouses, in all cases, are better for mushrooms than hothouses. Cool houses are seldom kept at

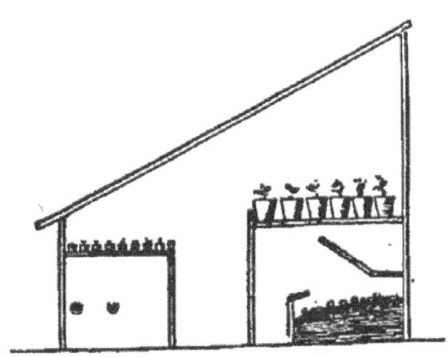

FIG. 11. BOXED MUSHROOM BED UNDER GREENHOUSE BENCH.

a lower temperature than 45° or 50° in winter, while hothouses run from 60° to 70° at night, with a rise of ten to twenty degrees by day, and this is too hot for mushrooms. It is a very easy matter, by means of cov-

41

ering with hay or boxing over and covering the boxing
with hay or matting, to keep a mushroom bed in a cool
house warm and free from marked changes in temperature; but it is a difficult matter to keep a mushroom
bed in a hothouse cool enough and prevent sudden rises
in temperature.

On Greenhouse Benches.—It sometimes happens
that the beds are formed on the greenhouse benches, and
the mushrooms occupy the same place that might be
assigned to roses or any other planted-out crop.   The
beds on the benches are made one board deep, that is,
eight to ten inches of short, fresh manure, and otherwise
as in the case of beds anywhere else.   After the beds
are spawned and cased with soil, by covering them over
with a layer of straw litter or hay, sudden drying out of
the surface is prevented, and in order to further prevent
this drying it is a good plan to sprinkle some water
over the mulching every day or two, but not enough
to soak through into the bed.   About the time the
young mushrooms commence to show themselves, remove
the mulching and replace it with a covering of shutters
raised another board's height above the bed, or with
strong calico or plant-protecting cloth hung curtain-
fashion over the beds.   The accompanying illustration,
Fig. 12, for which I am indebted to Henry A. Dreer, of
Philadelphia, gives an excellent idea of how mushrooms
may be grown and cared for on greenhouse benches.
This illustration, Mr. Dreer writes: "is made from a
photograph of a crop grown on the greenhouse benches
at the Model Farm, by Mr. McCaffrey, gardener to J. E.
Kingsley, Esq., of the Continental Hotel. . . . No
covering of litter is used, but the requisite shading on
sunny days is secured by the use of cotton cloth stretched
over the top of the bed, as shown in the engraving."

My principal objection to mushroom beds on greenhouse benches is their liability to frequent and marked

FIG. 12. MUSHROOMS GROWN ON GREENHOUSE BENCHES AT MR. J. E.
KINGSLEY'S MODEL FARM.

changes of atmospheric temperature and moisture, and to drying out. In midwinter they may be all right, but as spring advances and the sun's brightness and heat increase, the susceptibility of the beds to become dry also increases.

In Frames in the Greenhouses.—Mr. J. G. Gardner has a range of greenhouses some 900 feet long—the longest unbroken string of glasshouses that I know of—for the forcing of fruit and vegetables in winter; grapes, peaches, nectarines, figs, tomatoes, cucumbers, snap beans, peas, lettuce. This range is divided into several compartments, to accommodate the different varieties of crops, also so that some can be run as succession houses. In order to make the most of everything, market-gardener-like, he doubles up his crops wherever possible, and for this end he finds no crop more amenable and profitable than mushrooms. It

FIG. 13. WIDE BED WITH PATHWAY ABOVE.

matters nothing to him whether the house is cold or warm, he can grow mushrooms in it anyway, and in order to be master of the situation he makes his mushroom beds in hotbed frames inside the greenhouses. By attending to ventilating or keeping close, or covering up or leaving bare, he can properly regulate the temperature of the mushroom bed, no matter how hot or cold the atmosphere of the greenhouse may be. In the same way—by shading the panes or unshading them—he governs the light admitted to the mushrooms.

The greenhouses in which the mushrooms are grown are orchard houses, that is, glasshouses in which peach and nectarine trees are grown and forced. As these trees fruit and finish their growth early, it is necessary that they be kept as cool and inactive as possible in the fall and early winter, and started again into growth in late winter. In the fall, therefore, the fermenting material being confined in frames retains warmth enough for the proper development of the mushrooms, and as

FIG. 14. MUSHROOMS ON GREENHOUSE BENCHES UNDER TOMATOES.

the winter advances and the heat in the frames begins to wane it becomes necessary to begin heating the greenhouses in order to start the trees into bloom and growth, and thus are provided very favorable conditions for the continued production of the mushroom crop.

The frames used are common hotbed box frames seven

feet wide and carrying three and one-half feet wide
sashes.   A string of them is run along the middle of the
greenhouses, for greenhouse after greenhouse is occupied
by them.   They are flat upon the floor, and in the early
part of the season alone in the greenhouses.   But as the
winter advances a temporary staging is erected over
these frames, on which spiræas, peas, beans, or other
flowers or vegetables are to be grown.   These love the
light and a position near the glass, whereas the mushrooms grow perfectly well in the dark quarters of the
frames under the stages.   If he did not grow mushrooms
under these stages the room would be unoccupied, hence
unproductive ; but by occupying it with mushrooms he
not only gets peaches and snap beans at once out of the
same greenhouse, but also a crop of mushrooms, often
worth as much as the other two.

In preparing the beds in the frames they were made
up a foot deep, very firm, and with New York stable
manure brought direct from the cars.   There was no
preliminary preparation of the manure.   A layer of loam
one and one-half inches deep was then spread over the
surface and forked into the bed of manure one and one-
half inches deep, so as to form an earthy mat three
inches deep.   This was then packed solid with the feet,
and a two-inch layer of loose manure added all over.   In
about ten days the temperature three inches below the
surface was about 95°, and the beds were then spawned.
In spawning, drills were drawn across the beds about a
foot apart and just deep enough to touch but not penetrate the earthy mat before referred to.   The broken
spawn was then sown in the drills and covered with a
layer of loam one and one-half to two inches deep, which
was tamped slightly.   The sashes were then put on and
tilted up a little to let the moisture escape.   By the
time the mushrooms appeared there was very little need
of ventilating, as the condensation of moisture on the

glass was scarcely apparent; but ventilation is easily guided by the appearance of moisture on the glass, the more of this the more ventilation should be given. To begin with, there was no attempt at shading the frames; but as soon as the mushrooms began to appear the beds were shaded, and mostly by the crops of other plants on the stages above them. These frame beds were made up last October, and began bearing in December, and on March 14 Mr. Gardner wrote me: "The mushrooms in my frames have done grandly. I cut large basketfuls to-day of the finest mushrooms I have ever seen, some of them measuring five inches in diameter before being fully expanded."

And further, in submitting the above notes to him for verification, he adds: "There is one vital point we should impress upon all who grow mushrooms in frames or under greenhouse benches, namely, that sudden changes of temperature must be avoided. While light, in my opinion, is good for mushrooms, it causes a rise of temperature, and this we must guard against. In order to maintain a uniform temperature all glass exposed to light or heat in any other way should be covered with some non-conducting material. Rye straw is the best thing for this purpose that I know of. Indeed, neglect of this simple matter, in cases where sunlight and heat from hot-water pipes come in contact with the young mushrooms or mycelium on the surface of the beds, is the cause of many failures in growing in frames and greenhouses."

Under Greenhouse Benches.—Open empty spaces under the stages anywhere are good places for mushroom beds. However, carefully observe a few points, to wit: A dry floor under the beds is imperative, for a wet floor soaks and chills the beds, and renders them unhealthy for the spawn; but the common earth floor is good enough, provided water does not stand upon it at any time; if it does, the floor to be under the beds can be rendered dry by raising it a little higher than the general level, or using a flooring of old boards. Beds should not be built close up against hot-water pipes, steam pipes, or smoke flues, as the heat from these when they are in working condition will bake the parts of the beds next to them and render them unproductive, and also crack and spoil the caps of the mushrooms that come up within a foot or two of the pipes. But this injury from hot pipes and flues can be lessened greatly by boxing the pipes, so as to shut off the heat from the mushroom beds and allowing it full escape upward; then the beds can be made, with safety, up to within a foot of the pipes. As a rule, hot-water pipes are run around under the front benches of a greenhouse, then it would not be advisable to make beds under those benches. The middle bench is the one most commonly free from pipes, hence the one best adapted for beds. It has more headroom, and therefore easier working . facilities. Steam-heated greenhouses generally present the best accommodations for mushroom beds, because the pipes occupy less room under the benches than do those for hot water, and they are always kept higher from the ground.

Among Other Plants on Greenhouse Benches. —It sometimes happens that mushrooms spring up spontaneously among the roses, carnations, violets, mignonette, and other crops that are grown "planted out" on the benches, and this is particularly the case where fresh soil had just been used, in whole or part, for filling the bench beds. These mushrooms come from natural spawn contained in the loam or manure before they were brought indoors, and which is apt to be true virgin spawn. The mushrooms are generally of the common kind, grown from brick spawn, but occasionally a much larger and heavier sort is produced, and this is the "horse" mushroom. It is perfectly good to eat, only of coarser quality than the other.

A fair and certain crop can be obtained by planting pieces of spawn in the beds here and there between the plants and where they will be least likely to be soaked with water. In order to further insure the development of the spawn, holes about the size of a pint cup should be scooped out here and there over the bed, and filled up solidly with quite fresh but dry horse droppings, with the piece of spawn in the middle, and covered over on top with an inch of loam, so as to leave the whole surface of the bed level. So small a quantity of dry manure surrounded with cold earth will not heat perceptibly, and the moisture of the loam about it will soon moisten it, no matter how dry it may be. The dry, fresh droppings are the very best material for starting the mycelium into growth.

Growing Mushrooms in Rose Houses.—George Savage, the head gardener at Mr. Kimball's greenhouses, Rochester, N. Y., grows mushrooms very successfully under the benches of the rose houses. When he makes up his earliest mushroom beds in the fall the rose house is kept cool, and this is an advantage to the mushroom beds, which get all the warmth they need from the fermenting manure; but as November advances, and the heat in the beds begins to wane the rose houses are "started," and this artificial warmth comes in good season to benefit the growing mushrooms. The roses, in this case, are planted out on benches, hence there is scarcely any dripping of water from above upon the mushroom beds below.

Mr. George Grant, of Mamaroneck, N. Y., who grows mushrooms in the greenhouse, I called to see last January, and was very much pleased with his simple and successful method. The beds were then in fine bearing, very full, and the crop was of the best quality. The beds were made upon the earthen floor of his tomato-forcing house and under the back bench. The bed was

4

flat, seven to eight inches deep, with a casing of a ten-inch-wide hemlock board set on edge at the back, and another of same size against the front. The bed was made of horse droppings, six inches deep, and molded over with fresh loam one and one-half inch deep. Over the whole, and resting on the edges of the hemlock boards, was a light covering of other boards, with a sprinkling of hay on top of them to arrest and shed drip, and maintain an equable temperature in the bed.

Mr. Abram Van Siclen, of Jamaica, Long Island, is one of the largest mushroom growers for market in the country, as well as one of the most extensive growers of market-garden truck under glass around New York. He devotes an immense area under his lettuce-house benches to the cultivation of mushrooms. The beds are made upon the floor in the usual way, only for conven-ience' sake, to admit of plenty of room in making up the beds and gathering the crop, besides avoiding the necessity for building higher structures than the ordinary lettuce greenhouses, the mushroom beds are sunken about eighteen to twenty-four inches under the level of the pathways. As the lettuces are planted out upon the benches there is very little drip from them, hence the sunken beds are well enough. And the temperature of a lettuce house is about right for a long-lasting mushroom bed. Light is excluded by a simple covering of salt hay laid over the beds, and sometimes by light wooden shut-ters set up against the aperture between the lettuce benches and the floor, in this way boxing in the mush-rooms in total darkness.

Mr. William Wilson, of Astoria, has an immense greenhouse establishment near New York. In his green-houses, under both the side and middle benches, he grows mushrooms, and when I saw them in January there were about 300 square yards of beds. The beds were flat, about nine inches thick, built upon the ground, and protected from strong light by having muslin tacked over the openings between the benches and the beds alongside the pathways. But his crop was suffering from drip. Mr. Wilson told me he could not begin to

FIG. 15. MR. WM. WILSON'S MUSHROOM BEDS.

supply the demand. He says whatever he makes on mushrooms is mostly clear gain. They occupy space that otherwise would remain unoccupied, and he needs the manure and the loam in his florist business, and it is in better condition for potting after it has been rotted in the mushroom beds than it was before it was used for this purpose.

Drip from the Benches.—This must be prevented from the beds above, else it will soak or chill, and in a large measure kill the spawn. I have seen many examples of this evil. The beds would be full of drip holes all over their surface, and although a good many mushrooms here and there about the bed might perfect themselves, multitudes only reach the pin-head condition—or possibly the size of peas—and then fogg off in patches. It is not one or two little mushrooms in a clump that fogg off, but where one foggs off all of the little ones in

that patch go, for it is not a disease of the individual
mushroom, but of the mycelium or mushroom plant that
runs in the bed, and when this is injured or killed all
the little mushrooms arising from this particular patch
of plant are robbed of sustenance and must perish.

In greenhouses where the benches are occupied with
roses, carnations, bouvardias, violets, or lettuces, "planted out," as commercial florists and gardeners generally
grow them, there is very little drip, because while the
plants on these benches are freely watered, the soil is
never soaked enough for the water to drain from it in
dripping streamlets, as is continually the case in greenhouses where potted plants are grown on the stages.
Under these "planted out" benches, if care is exercised,
mushrooms can be grown in open beds; in fact, it is
about the best place and condition for them in a
greenhouse.

With stages occupied by plants in pots provision needs
to be made to ward off the drip from the mushroom beds,
by erecting over, and conveniently high above them,
a light wooden framework, on which rest light wooden
frames covered with oiled paper, oiled muslin, or plant-
protecting cloth.   In fact, three light wooden strips run

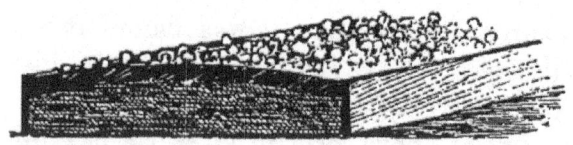

FIG. 16. MUSHROOM BED BUILT FLAT UPON THE GROUND.

over the bed, as shown in Fig. 12, or three strings of
stout cord or wire run in the same manner will answer
for small beds, and act as a support for the oiled muslin
or plant-protecting cloth.   Building paper is sometimes
used for the same purpose.   Mr. J. G. Gardner uses

ordinary hotbed frames and sashes, as described in a previous chapter. Light wooden shutters—made of one-half inch or five-eighths inch pine—may be used for the same end, and will last for many years.

The beds under the greenhouse benches may be made up in the same way as are beds anywhere else; that is,

FIG. 17. RIDGED MUSHROOM BED.

flat upon the floor and between two boards set on edge, as seen in Fig. 16, or in ridges under the high or middle benches, as in Fig. 17, or in banked beds against the

FIG. 18. BANKED BED AGAINST A WALL.

back wall, as shown in Fig. 18. Generally the flat bed is the most convenient to make and take care of.

In open, airy greenhouses it is always well to inclose the mushroom beds in box casings and with sash or shut-

ter coverings, to prevent draughts and fluctuations of
temperature and atmospheric moisture. This can easily
be done by making the sides a board and a half (fifteen
inches), or two boards (twenty inches) high, and covering over with light wooden shutters, sashes, or muslin
or paper-covered light frames. See Fig. 11.

Ammonia Arising.—Ammonia arising from the
manure of the mushroom beds in the greenhouse may be
injurious to the other inmates of the greenhouse. If
the manure has been well prepared before it was introduced into the greenhouse, the ammonia arising from
it will not, in the least degree, injure any other plants
or flowers that may be in the house; but if the manure
is fresh, hot, and rank, the opposite will be the case.
Beds in greenhouses should always be made up of manure
that has been well prepared beforehand out of doors or
in a shed, and as it is brought into the greenhouse it
should at once be built solidly into the beds. Then
very little steam will arise from the beds; in fact, it will
be imperceptible to sight or smell.

CHAPTER VI.

GROWING MUSHROOMS IN THE FIELDS.

Under suitable conditions we can grow mushrooms
easily and abundantly in the open fields, and the planting of the spawn is all the trouble they will cause us.
During the late summer and fall months mushrooms
often appear spontaneously and in great quantity in our
open pastures, but in their natural condition they are an
uncertain crop, as in one year they may occur in the
greatest abundance, and in the next perhaps none can be
found in the fields in which they had been so numerous

the previous year. Why this should be so is not very clear. The popular opinion is that after a dry summer mushrooms abound in the fields, but after a wet summer they are a very scarce crop; and the inference is that the moisture has killed the spawn in the ground. This may be true to a certain extent, but how does it happen —as it certainly often does—that good spawn planted by hand in the fields in early summer will produce mushrooms toward fall no matter whether the summer has been wet or dry? At the same time, it is true that a wet spell immediately succeeding the planting of the spawn will kill a great deal of it.

As a rule, wild mushrooms abound most in rich, old, well-drained, rolling pasture lands, and avoid dry, sandy, or wet places, or the neighborhood of trees and bushes. In attempting to cultivate them in the open fields we should endeavor to provide similar conditions. Then the chief requisite is good spawn, for without this we can not raise mushrooms.

About the middle of June take a sharp spade in the pasture, make **V** or **T**-shaped cuts in the grass sod about four inches deep and raise one side enough to allow the insertion of a bit of spawn two to three inches square under it, so that it shall be about two inches below the surface, then tamp the sod down. By cutting and raising the sod in this way, without breaking it off, it is not as likely to die of drought in summer. In this way plant as much or little as may be desired and at distances of three, four, or more feet apart. During the following August or September the mushrooms should show themselves, and continue in bearing for several weeks.

Mr. Henshaw, of Staten Island, who has been very successful in growing mushrooms in the fields as well as indoors, writes to me as follows: "You ask me to give you my plan of growing mushrooms in the fields during

the summer. It is very simple. About the end of
June, or as soon as dry weather sets in, we remove the
old beds from our mushroom house, and if there should
be any live spawn in the bottom of our beds we put it in
a wheelbarrow and take it to the field, where we plant it
in the open places, but never under trees. In planting,
we lift a sod and put a shovelful of the manure containing the spawn in the hole, then replace the sod and beat
it down firm; this we do at distances of twelve feet
apart. If we have no live spawn from our indoor beds
we take the common brick spawn, and put about a
quarter of a brick into each hole, returning and beating
down the sod as already stated. This is all that is done.
If there comes a dry time after the spawn is put in the
pasture we are sure to have a good supply of mushrooms
in the fall."

A few years ago Carter & Co., seedsmen, London, sent
this to one of the gardening periodicals : "The following mode of growing mushrooms in meadows by one of
our customers may be interesting to your readers : In
March (May would be soon enough here) he begins to
collect droppings from the stables. These, when enough
have been gathered together, are taken into the meadow,
where holes dug here and there about one foot or eighteen inches square are filled with them, the soil removed
being scattered over the surrounding grass. When all
the holes have been filled and made solid he then places
two or three pieces of spawn about one inch square in
each hole, treads all down firmly, replaces the turf and
beats it tightly down. Under this system, in August
and September mushrooms appear without fail in abundance and without any further care. The method is
simple and the result certain. Therefore all who happen
to have a meadow, paddock, or grass field, and are fond
of mushrooms, should try the experiment. . . . In
the case in question fresh holes were spawned every
year."

CHAPTER VII.

MANURE FOR MUSHROOM BEDS.

In order to grow mushrooms successfully and profitably a supply of fresh horse manure is needed, and this should be the very best that is made, either at home or bought from other stables. The questions of manure and spawn are the most important that we have to deal with. Very few make their own spawn, as it is bought and accepted upon its good looks,—often rather deceptive,—but the manure business is entirely in our own hands, and success with it depends absolutely upon ourselves. We can not reasonably expect good results from poor manure nor from ill-prepared manure. It is only from the very best of horse manure prepared in the very best fashion that we can hope for the very best crops of the best mushrooms.

Horse Manure.—There are various kinds of horse manure, differing materially in their worth for mushroom beds. The kind of manure depends upon the condition of the horses, how they are housed, fed, and bedded, and how the manure is taken care of. But while the manure of all healthy animals is useful for our purpose, there still is a great choice in horse manure. If we are dependent upon our home supply we may use and make the best of what we have, but if we have to buy the manure we should be very particular to select the best kind of manure and accept of no other.

The very best manure is that from strong, healthy, hard-worked, well-kept animals that are liberally fed with hard food, as timothy hay and grain, and bedded with straw. And if the bedding be pretty well wetted

57

with urine and trampled under the horses' feet, so much
the better; indeed, this is one reason why manure from
farm and teamsters' stables is better than that from

FIG. 19. PERSPECTIVE VIEW OF THE DOSORIS MUSHROOM CELLAR.

stylish establishments, where everything is kept so scru-
pulously dry and clean.

The fresher the manure is the better, still manure
that is not perfectly fresh may also be quite good.

Stable manure may accumulate in a cellar for a couple of months, and still be first rate. After our hotbed season is over I stack our stable manure high in the yard, and from June until August, as the manure is taken away from the stable each day, it is piled on the top of this stack. My object is to keep it so dry that it can neither heat nor rot. In August the stack is broken down and the best manure shaken out to one side for mushrooms, and the long straw and rotted parts thrown to the other side. This short manure, when moistened with water and thrown into a heap, exposed to the sun for a day or two, will heat up briskly. The beds illustrated in Fig. 19 were made from manure prepared in this way in August.

In the case of quite fresh manure, let it accumulate for a few days, or a fortnight, even, until there is enough of it to make up a bed, and then prepare it. Be very particular to prevent, from the first, its heating violently or "burning" while accumulating in the pile. Beds made from very fresh manure respond quickly and generously. The crop comes in heavily to begin with, and continues bearing largely while it lasts, but its duration is usually shorter than in the case of a bed made up of less fresh manure. But altogether it yields a better and heavier crop than a bed that comes in more gradually and lasts longer, and the mushrooms are of the finest quality.

Some growers use the droppings only, and reject all of the strawy part, or as much of it as they can conveniently shake out. This gives them an excellent manure and perhaps the very best for use on a small scale or in small beds. When mushrooms are to be grown in boxes, narrow troughs, half barrels, and other confined quarters, it is well to concentrate the manure as much as possible—use all the droppings and as little straw as you can. But droppings alone for large beds would take too much manure and cost too much, and they would not be any better than with a rougher manure.

Always preserve the wet, strawy part of the manure, along with the droppings, and mix and ferment them together, and in this way not only add largely to the bulk of the pile, but secure the benefits afforded by the urine without reducing, in any way, the strength or fermenting properties of the manure. Shake out all the rank, dry, strawy part of the manure and lay it aside for other purposes. This may be of further use as bedding in the stables, covering the mushroom beds after they have been made up, or for hotbeds; if well wetted with stable drainings, or even plain water, it forms a ready heating material.

Many a time when we have been short of home-made manure I have bought some loads here and there from different stables in the village, and mixed all together and made it into beds with excellent results. Sometimes when the manure under preparation had been rather old and cool, I have added a fifth or tenth part of fresh droppings to it, with very quickening effect in heating and apparent benefit to the crop.

It is generally believed that the manure of entire horses is better for mushrooms than that of other horses, but positive evidence in this direction has never come under my observation. Some practical men assert that there is no difference. Mr. John G. Gardner, at the Rancocas Farm, who has had abundant opportunity to test this matter, tells me that he has given it a fair trial and been unable to find any difference in the quality or quantity of mushrooms raised from beds made from the manure of entire horses and those raised from beds made from the manure of other equally as well fed animals. But the Parisian growers insist that there is a difference in favor of entire horses, especially in the case of hard-worked animals such as are engaged in heavy carting.

Manure of horses that are largely fed with carrots is emphatically condemned by most writers on the cultivation of mushrooms; indeed, it is one of *the* points in every book on mushrooms which I have read. Let us look at a few practical facts: There are at Dosoris two shelf beds in one cellar; each is thirty feet long, three feet wide, and nine inches deep, and both are bearing a very thick crop of mushrooms. The material in these beds consists of horse manure three parts and chopped sod loam one part, which had been mixed and fermented together from the first preparation. The manure was saved from the stables on the place in November, '88, the materials prepared in December, the beds built Dec. 17, spawned Dec. 24, molded over Dec. 31, and first mushrooms gathered Feb. 7, 1889. These beds bore well until the middle of April. The mushrooms did not average as large as they did on the deeper beds upon the floor of the cellar, but they ran about three-fourths to one ounce apiece, and a good many were more than this. It is most always the case, however, that the crop on thin shelf beds averages less than it does on thick floor beds, and especially is this noticeable after the first flush of the crop has been gathered, no matter what kind of fermenting material had been used. At the time when the manure used for these beds was being saved at the stable the horses were only very lightly worked, and to each horse was fed, in addition to hay and some oats and bran, about a third of a bushel of carrots a day. And this is the manure used for the late mushroom beds, and yet good crops and good mushrooms are produced. This is not only the experience of one year's practice but the regular routine of many.

Perhaps some one would like to ask: Do you consider the manure of carrot-fed horses as good as the manure of animals to which no carrots or other root crops had been fed? My answer is—decidedly not.

While the manure of carrot-fed animals is not the best,
at the same time it is good, and any one having plenty
of it can also have plenty of mushrooms. The complete
denunciation of the manure of carrot-fed horses so emphatically stereotyped upon the minds and pens of horticultural writers is not always founded on fact.

Manure of Mules.—This is regarded as being next
in value to that of entire horses, and some French
growers go so far as to say that it is quite as good. Mr.
John G. Gardner tells me of an extraordinary crop of
mushrooms he once had which astonished that veteran,
Samuel Henshaw, and that it was from beds made of
manure from mule stables. Certainly the heaviest crop
of mushrooms I ever did see was at Mr. Wilbur's place
at South Bethlehem, Pa., four years ago, and the beds
were of clean mule droppings from the coal mines.
Mule manure can be had in quantity at our mule stock
yards, which are in nearly every large city in the Middle
and Southern States. Getting it from the mines costs
more than it is worth, except as a fancy article; the
men will not collect and save it for any reasonable price.

Cellar Manure.—Many stables have cellars under
them into which the manure and urine are dropped at
every day's cleaning. These cellars are not generally
cleaned out before a good deal of manure has accumulated in them, say a few weeks', or a few months', or a
winter's gathering, and it is commonly pretty well moistened by the urine. If this manure has not become too
dry and "fire-fanged" in the cellar it is splendid for
mushrooms. We buy a good deal of it, but are particular to reject the very dry and white-burned parts.
Sometimes the manure from the cow-stables, as well as
from the horse-stables, is dropped together into the cellar; then I would give less for the manure, especially if
the cow manure predominated, because in the working
it keeps too cold and wet and pasty; but if there is not

cow manure enough to give the mass a pasty character
it will make capital mushroom beds. Pigs often have
the run of the manure-cellar, as is generally the case in
farmyards. I would not use any part of this mixed pig
manure. Mycelium evades hog manure; besides it is
impure and malodorous, and a propagating bed for noxious insect vermin. It matters very little what kind of
bedding is used, in the case of cellar manure, but I
would not buy it if sawdust or salt hay had been used as
bedding. Neither of these materials, in limited quantity, is deleterious to the mushrooms; at the same time,
they are far less desirable than straw, field hay, German
peat moss, or corn stalks, and there are risks enough in
mushroom-growing without courting any that we can as
well avoid.

City Stable Manure.—Around New York this can
always be had in any quantity at a reasonable rate, and
it is first-rate manure for mushroom beds. Market gardeners haul in a load of vegetables to market and bring
back a load of manure; others may buy and haul home
manure in the same way, or make arrangements with a
teamster to do it for them. But the whole matter of
city manure is now so deftly handled by agents, who
make a special business of it, that we can get any quantity of manure, from a 500 lb bale to an unlimited number of loads, and of most any quality, delivered near or
far, inland or coastwise, at a fairly moderate price. It
is the city stable manure that nearly all our large market growers use for their mushroom beds. When they
get it at the stables and cart it home themselves they
know what they are handling, and should take only fresh
horse dung. In ordering it of an agent be particular to
arrange for the freshest and cleanest, pure horse manure.
They will get it for you. We get several hundreds of
loads of this selected manure from them every year for
hotbeds, and find it excellent. We also get 1000 to 2000

loads of the common New York stable manure a year
for our general outdoor crops, and it also is capital
manure in its way, but not so good as the selected
manure for mushrooms.   It is mixed a little and smells
very rank, and in mushroom beds usually produces a
good deal of spurious fungi.   Most all of our largest
mushroom growers, Van Siclen of Jamaica, Denton of
Woodhaven, Connard of Hoboken, and others, live within
easy hauling distance of the city, and are able to select
and get the very choicest manure at a very cheap rate.

Baled Manure.—Within a year or two a good deal
of our city horse manure has been put up in bales and
thus shipped and sold.   Each bale contains from 350 to
nearly 500 lbs, and is made up, pressed and tied in about
the same way as baled hay.   The principal advantages
of the bales are these: Only the cleanest horse manure
is put up in this way; cow manure, offal, spent hops, or
other short or soft manures are not included in the bales,
nor, on account of shipping considerations, are malodorous manures of any sort permitted in them.   The rail-
roads allow baled manure to be put off on their platforms,
and closer to their stations than they would allow loose
manure; and it often happens that an agent will send a
carload to a railroad station and dump it off there so
that the people around who have only small garden lots
can have an opportunity of buying one or more bales,
just as they need it, and without, as is generally the
case, having to buy a whole load when they need only
half a load.   These bales are quite a boon to people who
would like to have a small bed of mushrooms in their
cellar and who have no other manure.   Bring home one
or more bales, open them, spread out the manure a little,
and when it heats turn it a few times, and it will soon
be ready for use.   Or if you do not wish to litter up the
place, roll the bales into the cellar, shed, or wherever
else you wish to make use of them, and mix about one-

fourth of their hulk of loam with the manure and make up the bed at once.

The Board of Health of New York city is very emphatic in its endeavors to rid the city of any accumulation of manure and, a year ago, had under consideration a plan to compel the manure agents, for sanitary reasons, to bale the stable manure. And perhaps this is the reason why it is so easily procured, to wit: A New York gentleman, desirous of engaging in the mushroom-growing business, writes me: "I get my manure from the city in bales. All it costs me is the freight to my place at White Plains." Lucky gentleman! With any amount of the best kind of stable manure gratis, no wonder he wishes to embark in the mushroom ship.

Cow Manure.—This is sometimes used with horse manure in forming the materials for a mushroom bed, and several European writers are emphatic in advocating its use. But I have tried it time and time again, and in various ways, and am satisfied that it has no advantage whatever over plain horse manure, if, indeed, it is as good. It is not used by the market growers in this country.

The best kind of cow manure is said to be the dry chips gathered from the open pastures; these are brought home, chopped up fine and mixed with horse manure. The time and expense incurred in collecting and chopping these "chips" completely overreach any advantages that might be derived from them, no matter how desirable they may be. The next best kind of cow manure is that of stall-fed cattle, to which dry food only, as hay and grain, is fed. This is seldom obtainable except in winter, and is then available for spring beds only. This I have used freely. One-third of it to two-thirds of dry horse manure works up very well, heats moderately, retains its warmth a long time, also its moisture without any tendency to pastiness; the mycelium travels through

5

it beautifully, and it bears fine mushrooms. Still, it is no better than plain horse manure. The poorest kind of cow manure is the fresh manure of cattle fed with green grass, ensilage, and root crops; indeed, such manure can not be used alone; it needs to be freely mixed with some absorbent, as dry loam, German moss, dry horse droppings, and the like, and even then I have utterly failed to perceive its advantages; it is a dirty mass to work, and quite cold.

In the manufacture of spawn, however, cow manure is a requisite ingredient, and here again the manure of dry fed animals is better than that of those fed with green and other soft food. But my chief objection to the use of cow manure in the mushroom beds is that it is a favorite breeding and feeding place for hosts of pernicious bugs and grubs and earth worms,—creatures that we had better repel.from, rather than encourage in, our mushroom beds.

German Peat Moss Stable Manure for Mushroom Beds.—Although I have not yet had an opportunity of trying this material for mushroom beds, Mr. Gardner, of Jobstown, has great faith in it; so, too, has that prince of English mushroom growers, Richard Gilbert, of Burghley, who relates his success with it in growing mushrooms in the English garden

FIG. 20. BALE OF GERMAN PEAT MOSS.

papers. This peat moss is a comparatively new thing in this country, and is used in place of straw for bedding horses. It is a great absorbent and soaks up much of the urine that, were straw used instead, would be likely to pass off into the drains. To this is ascribed its great virtue in mushroom culture. It should be mixed with loam when used for mushroom beds.

Sawdust Stable Manure for Mushroom Beds.—This is the manure obtained from stables where sawdust has been used for bedding for the horses. It is a good absorbent and retains considerable of the stable wettings. Such manure ferments well, makes up nicely into beds, the mycelium runs well in it, and good mushrooms are produced from it. But if I could get any other fairly good manure I wouldn't use it. I remember seeing it at Mr. Henshaw's place some years ago. He had bought a quantity of fresh stable manure from the Brighton coal yards, where sawdust had been used for bedding for the horses, and this he used for his mushroom beds. I went back again in a few months to see the bed in bearing, but it was not a success. At the same time, some European growers record great success with sawdust stable manure. George Bolas, Hopton, Wirkeworth, England, sent specimens of mushrooms that he grew on sawdust manure beds to the editor of the *Garden*, who pronounced them "in every way excellent." Mr. Bolas says : "In making up the bed I mixed about one-third of burnt earth with the sawdust, sand, and droppings. The mushrooms were longer in coming up than usual, the bed being in a close shed, without any heat whatever. They have, however, far exceeded my expectations."

Richard Gilbert, of Burghley, also wrote to the *Garden*, April 25, 1885 : "There is nothing new in growing mushrooms in sawdust. I have done it here for years past ; that is to say, after it had done service as a bed for horses, and got intermixed with their droppings. I have never been able to detect the least difference in size or quality between mushrooms grown in sawdust and those produced in the ordinary way."

Tree Leaves.—Forest tree leaves are often used for mushroom beds, sometimes alone, instead of manure, but more frequently mixed with horse manure to increase the bulk of the fermenting material. Oak tree leaves are the best ; quick-rotting leaves, like those of the

chestnut, maple, or linden, are not so good, and those
of coniferous trees are of no use whatever.   As the leaves
must be in a condition to heat readily they should be
fresh ; such are easily secured before winter sets in, but
in spring, after lying out under the winter's snow and
rain, their "vitality" is mostly gone.   But we can secure a large lot of dry leaves in the fall and pile them
where they will keep dry until required for use.   As
needed we can prepare a part of this pile by wetting the
leaves, taking them under cover to a warm south-facing
shed, and otherwise assisting fermentation just as if we
were preparing for a hotbed.   While moistening the
leaves with clean water will induce a good fermentation,
wetting them with liquid from the horse-stable urine
tanks will cause a brisk heat, and for mushrooms produce more genial conditions.

Mushroom beds composed in whole or part of fermenting tree leaves should be much deeper than would be
necessary were horse manure alone used ; for half leaves
and half manure, say fifteen inches deep ; for all leaves,
say twenty to thirty inches deep.

While mushroom spawn will run freely in leaf beds
and we can get good mushrooms from them, my experience has satisfied me that we do not get as fine crops
from these beds or any modification of them as from
the ordinary stable manure beds.   And we can not wonder much at this, considering that the wild mushroom
is scarcely ever found in the neighborhood of trees or
where leaf mold deposits occur.

Spent Hops.—We can make good use of this in one
way.   If we are short of good materials for a mushroom
bed, we can first make up the beds eight or ten inches
deep with fermenting spent hops, and above this lay a
four or five inch layer of horse manure, or this and loam
mixed.   The hops will keep up the warmth, and the
manure affords a congenial home for the mushroom

spawn. But we should never use spent hops alone, nor so near the surface of the beds that the spawn will have to travel through it.

Spent hops can be had for nothing, and our city brewers even pay a premium to the manure agents to take the hops away.

CHAPTER VIII.

PREPARATION OF THE MANURE.

Get as good a quality of fresh horse manure as you can, and in sufficient quantity for the amount of bed or beds you wish to make. Next get it into suitable condition for making up into beds. This can be done out of doors or under cover of a shed, but preferably in the shed. Out of doors the manure is under the drying influence of sun and wind, and it is also liable to become over-wetted by rain, but under cover we have full control of its condition. All the manure for beds between July and the end of October is prepared out of doors on a dry piece of ground, but what is used after the first of November, all through the winter, is handled in a shed open to the south. During the autumn months we get along very well with it out of doors; after every turning cover the heap with strawy litter to save it from the drying influences of sun and wind. Remove this covering when next turned, and lay light wooden shutters on top of it as a precaution against rain. In the shed in winter the manure is protected against rain and snow and we can always work it conveniently; when the shed is open to the south—as wagon and wood-sheds often are—we get the benefit of the warm sunshine in the daytime in starting fermentation in the manure, but

in the event of dull, cold weather, cover up the pile
quite snugly with straw and shutters to start the heat in
it.   Altogether, a warm, close shed would be better.

It seldom happens that one can get all the manure he
wants at one time ; it accumulates by degrees.   This is
the case with the market grower who uses many tons,
and hauls it home from the city stables a little at a
time ; also with the private grower, who uses only a few
bushels or half a cord, and has it accumulate for days or
weeks from his own stable.   As the manure accumulates
throw it into a pile, straw and all, but not into such a
big pile that it will heat violently ; and particularly
observe that it shall not "fire-fang" or "burn" in the
heap.   If it shows any tendency to do this, turn it over
loosely, sprinkle it freely with water, spread it out a
little, and after a few hours, or when it has cooled off
nicely, throw it up into a pile again and tread it firmly
to keep it moist and from heating hastily.

When enough manure has accumulated for a bed, prepare it in the following way :   Turn it over, shaking it
up loosely and mixing it all well together.   Throw aside
the dry, strawy part, also any white "burnt" manure
that may be in it, and all extraneous matter, as sticks,
stones, old tins, bones, leather straps, rags, scraps of
iron, or such other trash as we usually find in manure
heaps, but do not throw out any of the wet straw ; indeed, we should aim to retain all the straw that has
been well wetted in the stable.   If the manure is too
dry do not hesitate to sprinkle it freely with water, and
it will take a good deal of water to well moisten a heap
of dry manure.   Then throw it into a compact oblong
pile about three or four feet high, and tread it down a
little.   This is to prevent hasty and violent heating and
"burning," for firmly packed manure does not heat up
so readily or whiten so quickly as does a pile loosely
thrown together.   Leave it undisturbed until fermenta-

tion has started briskly, which in early fall may be in two or three days, or in winter in six to ten days, then turn it over again, shaking it up thoroughly and loosely and keeping what was outside before inside now, and what was inside before toward the outside now ; and if there are any unduly dry parts moisten them as you go along. Trim up the heap into the same shape as you had before, and again tread it down firmly. This compacting of the pile at every turning reduces the number of required turnings. When hot manure is turned and thrown loosely into a pile it regains its great heat so rapidly that it will need turning again within twenty-four hours, in order to save it from burning, and all practical men know that at every turning ammonia is wasted,—the most potent food of the mushroom. We should therefore endeavor to get along with as few turnings as possible ; at the same time, never allow any part of the manure to burn, even if we have to turn the heap every day. These turnings should be continued until the manure has lost its tendency to heat violently, and its hot, rank smell is gone,—usually in about three weeks' time. If the manure, or any part of it, is too dry at any turning, the dry part should be sprinkled with water and kept in the middle of the heap. Plain water is what is generally used for moistening the manure, but I sometimes use liquid from the stable tanks, which not only answers the purpose of wetting the dry materials, but it also is a powerful stimulant and welcome addition to the manure. But the greatest vigilance should be observed to guard against overmoistening the manure ; far better fail on the side of dryness than on that of wetness.

If the manure is too wet to begin with it should be spread out thinly and loosely and exposed to sun and wind, if practicable, to dry. Drying by exposure in this way is not as enervating as "burning" in a hot

pile, and better have recourse to any method of drying
the manure than use it wet. If, on account of the
weather or lack of convenience for drying, the manure
can not be dried enough, add dry loam, dry sand, dry
half-rotted leaves, dry peat moss, dry chaff, or dry finely
cut hay or straw, and mix together.

The proper condition of the manure, as regards dryness or moistness, can readily be known by handling it.
Take a handful of the manure and squeeze it tight ; it
should be unctuous enough to hold together in a lump,
and so dry that you can not squeeze a drop of water
out of it.

Some private gardeners in England lay particular
stress upon collecting the fresh droppings at the stables
every day, and spreading them out upon a shed or barn
floor to dry, and in this way keeping them dry and from
heating until enough has accumulated for a bed, when
the bed is made up entirely of this material, or of part
of this and part of loam. But market gardeners, the
ones whose bread and butter depend upon the crops
they raise, never practice this method, and that patri-
arch in the business, Richard Gilbert, denounces the
practice unstintedly.

Different growers have different ideas of preparing
manure for mushroom beds, but the aim of all is to get
it into the best possible condition with the least labor
and expense, and to guard against depriving it of any
more ammonia than can be helped. See Mr. Gardner's
method of preparing manure, p. 22.

Loam and Manure Mixed.—Mushroom beds are
often formed of loam and manure mixed together, say
one-third or one-fourth part of the whole being loam,
and the other two-thirds or three-fourths manure ; if a
larger proportion of loam is used it will render the beds
rather cold unless they are made unusually deep. I am
not prepared to affirm or deny that this mixed material

has any advantages over plain manure; I use it considerably every year and with good results; at the same time, I get as good crops from the plain manure beds. But it has many warm friends who are excellent growers.

In preparing this mixed material I use fresh sod loam well chopp d up, and add it to the manure in this way: First select the manure and throw it into a heap to ferment, as before explained; then after the first turning cover the heap with a layer of this loam about three or four inches thick, enough to arrest the steam; at the next turning mix this casing of loam with the manure, and when the heap is squared off add another coating of loam of the same thickness in the same way as before, and so on at each turning until the whole mass is fit for use, and the full complement of loam, say one-fourth the full bulk, has been added. In this way much of the ammonia that otherwise would be evaporated from the manure is arrested and retained.

Some growers, when they first shake out their fresh manure, add the full complement of loam to it at once and mix them together. Others, again, Mr. Denton, of Woodhaven, for instance, prepare the manure in the ordinary way, and when ready for use add the quota of loam. I use good sod loam for two reasons, namely, because it is the very best that can be used for the purpose, and, also, after being used in the mushroom beds it is a capital material, and in fine condition for use in potting soft-wooded plants. But the loam commonly used to mix with the manure is ordinary field soil. If the loam is ordinarily moist to begin with, and also the manure, there is very little likelihood of any of the material getting too dry during the preparation. And much less preparation is needed, for the presence of the loam lessens, considerably, the probability of hasty, violent fermentation.

Mr. Withington, of South Amboy, N. J., uses rather a stinted amount of loam in his manure. He writes me:

"We made up our beds this year with a proportion of loam in the manure, say one part loam to eight parts manure, but have always used clear manure heretofore, and I think the beds hold out longer than when only manure is used."

CHAPTER IX.

MAKING UP THE MUSHROOM BEDS.

The place in the cellar, shed, house, or elsewhere, where we intend to grow the mushrooms, should be in readiness as soon as the manure has been well prepared and is in proper condition for use. The bed or beds should be made up at once. The thickness of the beds depends a good deal upon circumstances, such as the quality of the manure,—whether it is plain horse manure, or manure and loam mixed together,—or whether the beds are to be made in heated or unheated buildings, and on the floor or on shelves. Floor beds are generally nine to fifteen inches deep; about nine inches in the case of manure alone, in warm quarters, and ten to fourteen inches when manure and loam are used. In cool houses the beds are made a few inches deeper than this so as to keep up a steady, mild warmth for a long time. The beds may be made flat, or ridged, or like a rounded bank against the wall; but the flat form is the commonest, and the most convenient where shelves are also used in the same building. Shelf beds are generally nine inches deep; that is, the depth of one board.

In making up the beds, bring in the manure and shake it up loosely and spread it evenly over the bed, beating it down firmly with the back of the fork as you go along, and continue in this way until the desired depth is attained. If it is a floor bed and there is no impediment, as a shelf overhead, tread the manure down firmly and evenly; if the manure is fairly dry and in good condition it will be pretty firm and still springy, but if it is too moist and poorly prepared treading will pack it together like wet rotten dung.

Now pierce a hole in the bed and insert a thermometer. There are "ground" or "bottom-heat" thermometers, as gardeners call them, for this purpose, but any common thermometer will do well enough; and after two or three days examine this thermometer daily to see what is the temperature of the manure in the bed. In roomy or airy structures or where only a small bed has been made it may, in the meantime, be left in this condition. But in a tight cellar I find that the warm moisture arising from the bed condenses in the atmosphere and settles on the top of the manure, making it perfectly wet. In order to counteract this, as soon as the bed is made up I spread some straw or hay over it loosely; the moisture settles on the covering and does not reach through to the manure. Beware of overcovering, as such induces overheating inside the bed. At spawning time remove this covering. The bed will then have become so cool (80° or 90°) that there is very little evaporation from it, consequently little danger of surface-wetting.

The Proper Temperature.—This, in mushroom beds, depends upon the materials of which they are composed, their thickness, how they are built, the situation they are in. and other circumstances. If the manure was good and fresh to begin with, carefully prepared and used as soon as ready, the bed in a few days will warm up to 125°, or a little more or less, and this is very good. My best beds have always shown a maximum heat of between 120° and 125°. Had the manure been used a few days too soon the heat would rise higher, perhaps to 135°, but this is too warm; in this case I

would fork over the surface of the bed a few inches deep
to let the heat escape, and after a couple of days compact the bed again. Boring holes all over the surface of
the beds with a crowbar is the common way of reducing
a too high temperature, and when the heat has subsided
sufficiently fill up these holes with finely pulverized dry
loam. With loam we can fill them up perfectly, but we
can not do this with manure, and if left open they remain as wet sweat holes that are very deleterious to the
spreading spawn.

A too high temperature in the beds should be sedu-
lously guarded against, for it wastes the substance of
the manure, dries up the interior of the bed, and the
mushroom crop must necessarily be starved and short.

Provided that the manure is fresh and good and has
been well prepared, if the beds, after being made up, do
not indicate more than 100° or 110° no alarm need be
felt, for excellent crops will likely be produced by these
beds. The thicker the beds are the higher the heat will
probably rise in them. Firmly built beds warm up
more slowly than do loosely built ones, and they keep
their heat longer. If the materials are quite cool when
built solidly into beds they are not apt to become very
warm afterward. But I always like to make up the
beds with moderately warm manure.

It sometimes happens that circumstances may prevent
the making up of the beds just as soon as the manure is
in prime condition, and even after they are made up the
heat does not rise above 75° or 80°. In such a case if
the manure is otherwise in good condition and fresh, it
is well enough and a good crop may be expected. But
if the manure, to begin with, had been a little stale,
rotten and inert, I certainly would not hesitate to at
once break up the bed, add some fresh horse droppings
to it, mix thoroughly, then make it up again. Or a fair
heat may be started in such a stale bed by sprinkling it

over rather freely with urine from the barnyard, then forking the surface over two or three inches deep and afterward compacting it slightly with the back of the fork. Spread a layer of hay, straw, or strawy stable litter a few inches deep over the bed till the heat rises. If the manure had been moist enough this sprinkling should not be resorted to, but the fresh droppings added instead. When it is applied, however, great care should be taken to prevent overheating; a lessening or entire removal of the strawy covering, and again firmly compacting the surface of the bed will reduce the temperature. Some saltpeter, or nitrate of soda, an ounce to three gallons of liquid, will encourage the spread of the mycelium after the spawn is inserted; a much stronger solution of these salts can now be used than would be safe to apply after the mycelium is running in the bed.

When loam and manure mixed together comprise the materials of which the bed is made, the temperature is not likely to rise so high as when manure alone is used, but this matters not so long as the materials of which the bed is composed are sweet and fresh and not over-moist. But if the materials are cold and stale treat as recommended for a manure bed, always bearing in mind that it is better to have a cold bed that is fairly dry than one that is wet, or, indeed, a warm one that is wet.

Mr. Withington, of South Amboy, has a good word to say for beds of a low temperature. He writes me: "Our beds kept in good bearing two months, though they have borne in a desultory way a month longer. Our best bed this season was one that was kept at an even temperature. The manure never rose above 75° when made up, and decreased to about 60° soon after spawning. Kept the house at 55°."

CHAPTER X.

What is mushroom spawn? Is it a seed or a root? Do you plant it or sow it, or how do you prepare it? are some of the questions asked me now and again. To the general public there seems to be some great mystery surrounding this spawn question; in fact, it appears to be the chief enigma connected with mushroom-growing. Now, the truth is, there is no mystery at all about the matter. What practical mushroom growers call spawn, botanists term mycelium.

The spawn is the true mushroom plant and permeates the ground, manure, or other material in which it may be growing; and what we know as mushrooms is the fruit of the mushroom plant. The spawn is represented by a delicate white mold-like network of whitish threads which traverse the soil or manure. Under favorable circumstances it grows and spreads rapidly, and in due time produces fruit, or mushrooms as we call them. The mushrooms bear myriads of spores which are analogous to seeds, and these spores become diffused in the atmosphere and fall upon the ground. It is reasonable to suppose that they are the origin of the spawn which produces the natural mushrooms in the fields, also the spawn we find in manure heaps. But we never have been able to produce spawn artificially from spores, or in other words, mushrooms have never been grown by man, so far as I can find any authentic record, from "seed." How, then, do we get the spawn? By propagation by division. We take the mushroom plant or spawn, as we call it, and break it up into pieces, and plant these pieces separately in a prepared bed of manure or other material, under conditions favorable for their growth, and we find that these pieces of spawn develop into vigorous plants that bear fruit (mushrooms) in about two months from planting time. When the spawn has borne its full crop of fruit it dies.

Well, then, if we can not produce spawn from spores, and the spawn in the beds that have borne mushrooms has died out, how are we to get the spawn for our future crops? is a question that may suggest itself to the inexperienced. By securing it when it is in its most vigorous condition, which is before it begins to show signs of forming mushrooms, and drying it, and keeping it dry till required for use. But in order to secure the spawn we need to take and keep with it the manure to which it adheres or in which it is spreading. In this way it can be kept in good condition for several years and without its vitality being perceptibly impaired. Keeping it dry merely suspends its growth; as soon as it is again submitted to favorable conditions of moisture and heat its pristine activity returns.

Mushroom spawn can be obtained at any seed store. Our seedsmen always keep it in stock, both the brick (English), and the flake (French) spawn. It is retailed in quantities of one pound or more, and as the article is perfectly dry it can be easily sent by mail in small quantities.

The seedsmen import it from Europe every year along with their seeds. A prominent Boston seedsman writes me: "We get our supply through the London wholesale seedsmen, for the sake of convenience and cheaper ocean freight, etc. Coming with a shipment of other goods and on same bill of lading brings the freight charges down. The low price at which mushroom spawn is sold in quantity can only be maintained with low freight rates, as there is a duty here of 20% on the article."

By direct inquiry of the leading importers in different cities I find that we import about 4500 lbs of French or flake spawn, and 4000 bushels, or 64,000 lbs of English or brick spawn, and that fully a half of this whole importation is handled by the seedsmen of New York city. In New York one firm alone, who make a specialty of supplying market gardeners, has in one year imported 1500 bushels of brick spawn.   But the vicinity of New

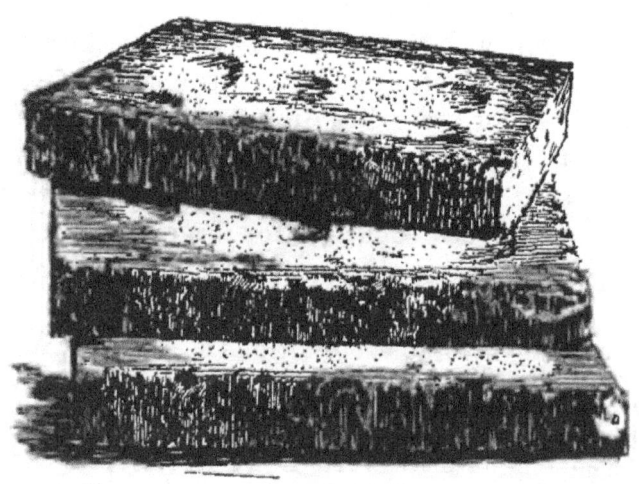

FIG. 21. BRICK SPAWN.

York is the great mushroom-growing center of the country, also the best market for mushrooms in the country.   One gardener at Jamaica, L. I., bought 1000 lbs of brick spawn at one time, and a neighbor of his bought 400 lbs; this shows what a large quantity of spawn market gardeners require.   And the demand this year is unprecedented; some of our leading importers had sold out their supply before the first of November. And it is not private growers so much as market growers

who are the cause of this; the market men find there is
money in growing mushrooms and they are going into it.

Spawn comes in the form of dry, hard, solid manure
bricks, and also in the form of flakes of half rotted
strawy manure. These bricks and flakes are completely
permeated with the mushroom mycelium.

The brick spawn is commonly known as English
spawn, and what is imported into this country is made
in England, mostly about London. The bricks made by
the different manufacturers vary a little in size and
weight; in some cases ten bricks go to the bushel, in
others fourteen, and in others sixteen. This last is the
commonest sized brick, and weighs exactly a pound, and
measures about eight and one-half inches long, five and
one-fourth inches wide, and one and one-fourth inches
thick; it is what the London spawn makers call a 9x6x2
inch brick, but it shrinks in drying. In retailing brick
spawn in this country it is sold by weight and not by
measure.

Mill-track mushroom spawn is advertised by some of
our seedsmen, but what they sell under this name is
only the ordinary English brick spawn. One of our
prominent seed firms who advertise it write me: "Genuine mill-track spawn used to be the best in England,
but it has been superseded, although European gardeners still call for English spawn under the name of 'mill-
track.'" The real mill-track spawn is the natural spawn
that has spread through the thoroughly amalgamated
horse droppings in mill-tracks or the cleanings from
mill-tracks. It is usually sold in large, irregular, somewhat soft lumps, and is much esteemed by spawn makers
for impregnating their bricks, but nowadays, that horses
have given place to steam as a motive power in mills, we
have no further supply of mill-track spawn for use in
spawning our mushroom beds. We do not feel this loss,
however, as the spawn now manufactured by our best

6

makers will produce as good a crop of mushrooms as the
old mill-track natural spawn used to do.

The flake spawn is what is generally known as French
spawn, and is imported into this country from France.
But the manufacture of "French" spawn for sale, however, is not strictly confined to France. It is put up in

FIG. 22. FLAKE OR FRENCH SPAWN.

two ways, namely, nicely packed in thin wooden boxes,
each containing two or three pounds of spawn, and also
loose in bulk when it is sold by weight or measure.

Virgin spawn is what we call natural spawn or wild
spawn; that is, the spawn that occurs naturally in the
fields, in manure piles, or elsewhere, and without any
artificial aid. It is supposed to be produced directly
from the mushroom spores, and is not a new growth of
surviving parts of old spawn that may have lived over in
the ground. It is far more vigorous than "made"
spawn, and spawn makers always endeavor to get it to
use in spawning the artificial spawn. It is seldom used
for spawning mushroom beds because not easy to obtain.
Now and again we come upon a lot of it in a manure
pile; it looks like a netted mass of white strings travers-
ing the manure. As soon as discovered secure all you
can find, bring it indoors to a loft, shed, or room, and
spread it out to dry; after drying it thoroughly keep it

dry and preserve and use it as you would French spawn, for it is the best kind of flake spawn. In using virgin spawn for spawning beds I have obtained larger and heavier mushrooms than from "made" spawn, and the beds lasted longer in good bearing, but the weight of the whole crop has not been more than from artificial spawn.

How to Keep Spawn.—Spawn should be kept in a dry, airy place, somewhat dark, if convenient, and in a temperature between 35° and 65°. Wherever things will "must," as in a cellar, cupboard against a wall, or in a close, damp building, is a very poor place for keeping spawn. If the spawn is perfectly dry and kept in a dry, airy place, and not in large bulk, and covered, it will bear a high temperature with apparent impunity, but whenever dampness, even of the atmosphere, is coupled with heat, the mycelium begins to grow, and this, in the storeroom, is ruinous to the spawn. Judging from our natural-mushroom crops, the spawn for which must be alive in the ground in winter, one concludes that frost should not be injurious to the artificial spawn, still my experience is that hard frost destroys the vitality of both brick and flake spawn. And this is one reason why I get our full supply of spawn in the fall and keep it myself rather than submit it to the mercy of the seed store.

New Versus Old Spawn.—How long spawn may be kept without its vitality becoming impaired is an unsettled question, but there is no doubt, if properly kept, it will remain good for several years. But I can not impress too strongly upon the reader the importance of using fresh spawn. Do not use any old spawn at any price ; do not accept it gratis and ruin your prospect of success by using it. It takes three months from the time when the manure is gathered for the beds until the mushrooms are harvested. Can you, therefore, afford to spend this time, and undergo the care and trouble and expense, and court a failure by using old spawn? We have risks enough with new spawn, let alone old spawn. I do not use any more old spawn, but I have used it often and long enough to be convinced of its general worthlessness, unless preserved with the greatest care.

How to Distinguish Good from Poor Spawn.— This is a very difficult matter, notwithstanding what people may say to the contrary. If we could positively tell good from bad spawn, we would never use bad spawn, and, therefore, with ordinary care, have very few failures in mushroom-growing; for good spawn is the root of success in this business. Spawn differs very much in its appearance; sometimes the bricks show very little appearance of the presence of spawn, and still are perfectly good; and again, we may get bricks that are pretty well interlaced and clouded with bluish white mold or fine threads, and this, too, is good. When the bricks are freely pervaded with pronounced white threads this is no sign that the spawn is bad. Bricks dried as hard as a board may be perfectly good; so, too, may be those that are comparatively soft. Mushroom spawn should have a decided smell of mushrooms, and whatever cobweb-like mold may be apparent should be of a fresh bluish white color, and the fine threads clear white. Prominent yellowish threads or veins are a sign that the mycelium had started to grow and been killed. Distinct white mold patches on the surface of the bricks indicate the presence of some other fungous parasite on the mushroom mycelium; the absence of any mushroom smell in the spawn indicates its worthlessness and that the mycelium is dead. One familiar with mushroom spawn can tell with considerable certainty "very living" spawn and "very dead" spawn, but I am far from convinced that any one can decide unhesitatingly in the case of middling or weak spawn.

Mr. S. Henshaw, in Henderson's Hand-book of Plants, tells us: "The quality of the spawn may be very easily detected by the mushroom-like smell, . . . and I should have no hesitation in picking out good spawn in the dark." Sanguine, surely, but I have tried it and found the test wanting. M. Lachaume says that good spawn shows "an abundance of bluish-white filaments well fitted together, and giving off a strongly marked odor of mushrooms. All those portions which show traces of white or yellow mold or have a floury appearance, should be rejected and destroyed." Mr. Wright says: "A brick may be a mass of moldiness, and yet be quite worthless; and if the mold has a spotted appearance, as if fine white sand had been dredged on and through the mass, it is certain there is no mushroom-growing power there. . . . If thick threads pass through the mass and there are signs of miniature tubercles on them, then the spawn may be regarded as too far gone. . . . Clusters of white specks on the spawn denote sterility."

Mr. A. D. Cowan, of New York, who has the reputation of being an excellent judge of mushroom spawn, writes me: "To correctly judge the quality of brick spawn by its appearance requires experience in handling it, and a trained eye which enables one quickly to detect good from bad, fair to middling. As two lots seldom come exactly or nearly alike in appearance, it is hardly possible to give precise rules to follow, excepting the never-failing requisite which the spawn must possess to be good, namely, the moldy appearance on the surface, the more the better, without showing threads. Too many of these to a given space are a sure indication of exhausted vitality, arising generally from the bricks being heaped together when in process of manufacture, before they are sufficiently dried. Healthy bricks are usually of a dusty brown color, and of light weight.

Black colored spawn is to be avoided, as a rule, and when the black appearance is very prevalent in a cargo of bricks it is a strong indication that the spawn has not run its course; and as it is not expected to do so after it has reached the hands of the retailer it is economy to cast it aside. Some persons break a brick into several pieces to see how it looks inside. To the experienced eye this is not necessary, or even to lay hands upon it, as the outward moldy appearance is the best of all evidence of its healthy vitality, and this never exists if the bricks have lost their germinating power, excepting, of course, where they have been kept damp, and the spawn has spent its power, which is detected by the white threads appearing in great quantity."

American-made Spawn.—So far as I have been able to find out by diligent inquiry, mushroom spawn is not made for sale in this country. But I am informed that a few growers do save and use their own flake spawn. Some of our principal growers, Van Siclen, Gardner, and Henshaw, for instance, in time past attempted to make their own spawn, but with only partial success, and now they confine themselves to the imported article. But this state of affairs can not long continue. The demand here for fresh mushrooms is so great, the industry of mushroom-growing so important, the price of imported spawn so high, and the quantity of foreign spawn imported annually into this country is so large, that, before long, we hope some one will find it to his advantage to make a specialty of growing mushroom spawn in this country to supply the American market. There is no practical operation in connection with the cultivation of mushrooms so little known or understood by the general grower as the growing (or "making," as it is commonly called) and preserving of mushroom spawn. General cultivators in England and France (outside of the Paris caves) do not make their own spawn; it is a distinct branch of the business, and carried on by specialists who grow mushrooms for sale in winter, and spawn in summer.

The time and attention required to produce a small quantity of first-class spawn are worth more than the cost of the spawn at the seed store. In order to make spawn profitably we must make it in large quantity, and we need not attempt to make it unless we have good materials and conditions for its proper preparation, and will give it every attention possible for its best development.

Because spawn may be made in America is no reason whatever why the American people will buy it. We must produce, at least, as good an article as the best in Europe before we can find countenance in our home market. It is not the shape of the manure brick, its size, fine finish, hardness, softness, or freshness, that counts in this case; it is the fullness and vitality of the mass of mycelium or mushroom plant that is contained within it.

HOW TO MAKE BRICK SPAWN.

As the making of brick spawn for sale is not yet an American industry, but almost entirely confined to England, I think it best to restrict myself to describing how it is made in England. Mr. John F. Barter, of Lancefield street, London, is one of the most successful mushroom growers and spawn makers in Great Britain. He writes me that he confines himself entirely to the mushroom business; he makes his living by it. He grows mushrooms in the winter months and makes spawn in the summer months; he employs men for mushroom bed making from August until March, then, to keep on the same hands during summer, he makes spawn for sale. He grows for and sells in the London market about 21,000 pounds of mushrooms a year, and in summer makes some 10,000 bushels, equal to 160,000 pounds, of brick spawn for sale. The amount of spawn made in a year by this one manufacturer is about three times as much as the total annual importation of mushroom spawn of all kinds into this country. And he is only one maker among several. This fact alone must convince us that mushroom-growing is carried on to a vastly greater extent in European countries than it is here, where we have as good facilities as they have, and an immensely better market.

The manner of making the spawn differs a little with the different manufacturers, and no one can become proficient in it without practical knowledge. I asked Mr. Barter if he thought spawn could be made profitably in this country, paying, as we do, $1.50 a day for laborers, and without any certainty of the same men staying with us permanently. He writes me: "Uncertain labor would be of no use. Of course the wages you pay would not affect it much, as I pay nearly as much as that for my leading men. But to begin with, you must have a man that has had some experience."

About the simplest and best way of making brick spawn that I find described is the following from *The Gardeners' Assistant*. I may here state that Robert Thompson, the author of this work, was for many years the superintendent of the Royal Horticultural Society's gardens at Chiswick, near London, and, in his day, was regarded as without a peer in practical horticulture, and lived in the midst of the market gardens of London and the principal mushroom-growing district.

"Fresh horse droppings, cow dung, and a little loam mixed and beaten up with as much stable drainings as may be necessary to reduce the whole to the consistence of mortar. It may then be spread on the floor of an open shed, and when somewhat firm it may be cut into cakes of six inches square. These should be placed on edge in a dry, airy place, and must be frequently turned and protected from rain. When half dry make a hole in the broadside of each, large enough to admit of about an inch square of good old spawn being inserted so deep as to be a little below the surface ; close it with some moist material the same as used in making the bricks. When the bricks are nearly dry make, on a dry bottom, a layer nine inches thick of horse dung prepared as for a hotbed, and on this pile the bricks rather openly. Cover with litter so that the steam and heat of the layer of dung may circulate among the bricks. The temperature, however, should not rise above 60° ; therefore, if it is likely to do so, the covering must be reduced accordingly. The spawn will soon begin to run through the bricks, which should be frequently examined whilst the process of spawning is going on, and when, on breaking, the spawn appears throughout pretty abundantly, like a white mold, the process has gone far enough. If allowed to proceed the spawn would form threads and small tubercles, which is a stage too far advanced for the retention of its vegetative powers. Therefore, when the spawn is observed to pervade the bricks throughout like a white mold, and before it assumes the thread-like form, it should be removed and allowed to dry in order to arrest the further progress of vegetation till required for use. It ought to be kept in a dark and perfectly dry place." I would add, do not keep it where it is apt to become musty or moldy in summer; also keep it in as cool a dry place as possible in summer, and always above 35° in winter.

These other recipes are also given :

"1. Horse droppings one part, cow dung one-fourth, loam one twentieth.

"2. Fresh horse droppings mixed with short litter one part, cow dung one-third, and a small portion of loam.

"3. Equal parts of horse dung, cow dung, and sheep's dung, with the addition of some rotten leaves or old hotbed dung.

"4. Horse dung one part, cow dung two parts, sheep's dung one part.

"5. Horse droppings from the roads one part, cow dung two parts, mixed with a little loam.

"6. Horse dung, cow dung, and loam, in equal parts."

From the above it appears that horse dung and cow dung are the principals in spawn bricks; the loam is added for the purpose of making the other materials hold together; it also absorbs the ammonia, which otherwise would pass off.

J. Burton's Method. From *The Kitchen and Market Garden.*—Make the spawn in early spring. As cow manure is the principal ingredient used in making the bricks this should be secured before the animals get any green food. Store it on the floor of an open, dry, airy shed, and turn it every few days for a week or two. Then add an equal part of the following: Fresh horse droppings, a little loam, and chopped straw, mixed together. "The whole should then be worked well together and then trodden down, after which it may be allowed to remain for a few days, when it will be required to be turned two or three times a week. If the weather be fine and dry the mass will soon be in a fit condition for molding into bricks, which process can be performed by using a mold in the same way as the brick makers, or, . . . the manure may be spread evenly on the floor to a thickness of six inches, and then be firmly trodden and beaten down evenly with the back of the spade. It should then be lined out to the required size of the bricks, and be cut with a sharp spade or turfing iron. In a few days the bricks will be sufficiently dry to handle, when they should be set up edgeways to dry thoroughly, and if exposed to the sun for two or three days they will be ready to receive the spawn. In introducing the spawn two holes large enough to admit a piece of spawn as big as a pigeon's egg should be cut in each brick at equal distances. This should be well beaten in and the surface made even with a little manure. The bricks should then be collected together in a heap and covered with enough short manure to cause a gentle heat, being careful that there is no rank heat or steam to kill the spawn. This must be carefully attended to until the spawn is found to have penetrated through the whole of the bricks, after which they should be stacked away in any convenient dry place."

HOW TO MAKE FRENCH (flake) SPAWN.

I can not do better than to let a practical Frenchman engaged in the business tell this story. In Vol. XIII of the London *Garden* I find an English translation of M. Lachaume's book, "The Cave Mushroom," and this comment by the editor: "The most complete account of the cave culture of mushrooms which has been published by any cultivator on the spot well acquainted with the subject is that recently published by M. Lachaume."

Lachaume says: "The best spawn to use is what is called 'virgin spawn'; that is to say, which has not yet produced mushrooms. In this country this kind of spawn may be procured of any respectable nurseryman, under the name of 'French spawn.' It differs from English spawn by being in the form of small tufty cakes, instead of in compact blocks. Large mushroom growers, however, always provide themselves with their own spawn by taking it from a bed which is just about to produce its crop, or which has already produced a few small mushrooms. . . . It is true that by thus 'breeding in and in,' as it were, the mushrooms show a tendency to deteriorate after a time; new spawn must therefore be obtained as soon as any signs of deterioration begin to manifest themselves."

Making French Virgin Spawn.—Condensed from Lachaume's book on mushrooms. Take five or six barrow loads of horse droppings that have lain in a heap for some time, and lost their heat, and mix them with one-fourth of their bulk of short stable litter. Then, in April, open a trench two feet wide, twenty inches deep, and length to suit, at the foot of, but eight inches distant from, a wall facing north. In the bottom of the trench spread a layer three to four inches deep of chopped straw, then an equally thick layer of the prepared manure, all pressed firmly by treading it down. The two layers must now be gently watered, and then another double layer of chopped straw and droppings must be laid, trodden down and watered, and so on until the top of the trench is reached. The bed ought to rise above the level of the ground and be rounded off like the top of a trunk. To prevent excessive dampness from heavy rain cover the mound with a thick layer of stable litter. Three months after filling the trench it should be opened at the side or end. If the pieces of manure are well covered with masses of bluish-white filaments, giving off the odor of mushrooms, the operation has succeeded, and the spawn is fit for use or for drying to preserve for future use. But if the threads are only sparingly scattered through the mass, the trench should be covered up again and left for another month. In saving the spawn the flakes of manure containing the largest amount of spawn filaments should be retained, and those showing a brown appearance rejected. In order to facilitate the drying of the spawn the flakes should be broken into pieces, weighing from one to two pounds; they are then placed in a well ventilated shed, but they must not be piled upon each other. Properly prepared and dried this spawn keeps good for ten years.

A Second Method (by Lachaume). "This is generally adopted by mushroom growers. The formation of
the spawn is accelerated by adding pieces of old spawn
here and there. . . . At the beginning of April we
must choose a piece of ground situated at the foot of a
wall facing north. . . . The soil ought to be very
open and light rather than heavy, so as to avoid dampness. Taking advantage of a fine day, we open a trench
sixteen inches wide and at about eight inches from the
foot of the wall, and of a length adapted to the quantity
of spawn we desire to produce. The earth is thrown
out on the side opposite the wall. Manure which has
been prepared for a mushroom bed, and has just come
into condition is then filled into the trench, leaving,
however, a space at one end of it about two feet and six
inches in length for the formation of a mushroom bed,
which is made by tossing the manure about and shaking
it up with the hands, after which it is pressed down
with the hands and-knees. As soon as the layer of
manure reaches six inches in thickness we place along
the edge a number of lumps of spawn at about one foot
apart. These lumps are placed level with the manure
on the edge facing the wall. This portion of the surface
of the manure ought to be raised vertically, and should
lean against the earthen wall of the trench. The other
half of the surface ought to slope gently toward the
wall, leaving a space of three or four inches between it
and the side of the trench, so that it may be trimmed.
The lumps of spawn on this surface should be placed a
little backward, so that they may not be broken when
the bed is trimmed. The bed is then covered with more
manure, until the first lumps of spawn are buried three
or four inches deep. A second row of lumps of spawn
is then inserted, as described in the directions for making the first row, and the bed is filled up level with the
surface of the soil. It is finished by covering it up with

a layer of fine, dry soil three or four inches thick. The
spawn ought to be very dry, otherwise we shall get a
premature crop of mushrooms instead of fresh spawn.
At the end of six weeks or a couple of months the new
spawn ought to make its appearance, a fact which we
may learn by opening the bed. One sign, which will
save us the trouble of opening up the beds, is the appearance of young mushrooms on the surface. The layer of
earth is first removed, and then the cakes of spawn are
treated as described in the directions given for the first
method of making spawn."

Third Method (by Lachaume). "By filling in a
trench like that described in the first method, by a series
of layers of one-third of pigeon or fowl guano, and two-
thirds of short manure, containing a large proportion of
spent horse droppings, treading it down firmly, watering
it if it is too dry, and finishing up with a layer of soil,
as described already, we may, at the end of a couple of
months, or even a little longer, procure a supply of well-
formed cakes of spawn of excellent quality, which may
be used in the ordinary manner."

From Mr. Robinson's "Mushroom Culture." "This
(French) spawn is obtained by preparing a little bed, as
if for mushrooms, in the ordinary way, and spawning it
with morsels of virgin spawn, if that is obtainable ; and
then when the spawn has spread through it, the bed is
broken up and used for spawning beds in the caves, or
dried and preserved for sale."

From Mr. Wright's book on mushrooms. "French
spawn . . is contained in flakes of manure. Neither
is it virgin spawn, nor derived immediately from it,
. . . but is spawn taken from one bed for impregnating another."

Relative Merits of Flake and Brick Spawn.—
The flake or French spawn costs about three times as
much as the brick or English spawn, and, as it is so

much whiter with mycelium than is the brick spawn,
many believe that it is more potent and well worth the
additional cost. In spawning the beds I use two pounds
of flake spawn to plant the same space for which I would
use five pounds of brick spawn, and this gives a capital
crop, with number of mushrooms a little in favor of the
flake spawn, but on account of the larger size of the
mushrooms the weight of crop is considerably in favor of
the brick spawn. And I find more certainty of a crop
in the case of the brick spawn than in the other.

Regarding the respective merits of brick and flake
spawn, Mr. Barter, in response to my inquiry, writes
me : "I have tried them both, and know brick spawn
to be far the best. You see, I do nothing but this
mushroom business for a living, so, of course, would
use the best kind of spawn for my crop. Generally the
French spawn produces one-third less mushrooms than
does the brick spawn from the same length of bed,
besides, those from the brick spawn are by far the heaviest and fleshiest."

I would here observe that Mr. Barter's remarks apply
more to ridge beds out of doors than beds in the cellar
or mushroom house. And it is odd, but true, that the
flake spawn does not produce as good results in outdoor
beds as it does in those under cover.

CHAPTER XI.

After the mushroom bed is made up it should, within a few days, warm to a temperature of 110° to 120°. Carefully observe this, and never spawn a bed when the heat is rising, or when it is warmer than 100°, but always when it is on the decline and under 90°. In this there is perfect safety. Have a ground thermometer and keep it plunged in the bed; by pulling it out and looking at it one can know exactly the temperature of the bed. Have a few straight, smooth stakes, like short walking canes, and stick the end of these into the bed, twelve to twenty feet apart; by pulling them out and feeling them with the hand one can tell pretty closely what the temperature of the bed is.

All practical mushroom growers know that if the temperature of a twelve inch thick bed at seven inches from the surface is 100°, that within an inch of the surface of the bed will only be about 95° indoors, and 85° to 90° out of doors. Also, that when the heat of the manure is on the decline it falls quite rapidly, five, often ten degrees, a day, till it reaches about 75°, and between that and 65° it may rest for weeks.

Some years ago I gave considerable attention to this matter of spawning beds at different temperatures. Spawn planted as soon as the bed was made (five days after spawning the heat in interior of bed ran up to 123°) yielded no mushrooms, the mycelium being killed. The same was the case in all beds where the spawn had been planted before the heat in the beds had attained its max-

imum (120° or over). Where the heat in the middle of
the bed never reached 115°, the spawn put in when the
bed was made, and molded over the same day, yielded a
small crop of mushrooms. A bed in which the heat was
declining was spawned at 110°; this bore a very good
crop, and at 100° and under to 65° good crops in every
case were secured, with several days' delay in bearing in
the case of the lowest temperatures. But notwithstanding these facts, my advice to all beginners in mushroom
growing is, wait until the heat of the bed is on the decline
and fallen to at least 90°, before inserting the spawn.

Writing to me about spawning his beds, Mr. Withington, of New Jersey, says: "I believe a bed spawned
at 60° to 70°, and kept at 55° after the mushrooms
appear, will give better results than one spawned at a
higher temperature, say 90°."

Preparing the Spawn.—If brick spawn is used cut
up the bricks (standard size) into ten or twelve pieces
with a sharp hatchet, and avoid, as much as possible,

BRICK SPAWN CUT IN PIECES FOR PLANTING.

making many crumbs, as is the case generally when a
hammer or mallet is used in breaking the bricks. Extra
large pieces of spawn are apt to produce large clumps of
mushrooms, but this is not always an advantage, as
when many mushrooms grow together in a clump they
are apt to be somewhat undersized, and in gathering we
can not pluck them all out clean enough so as not to
leave a part of the "root" in the ground to poison the

7

balance of the clump, in cases where several or many of
them spring from one common base.

Inserting the Spawn.—When brick spawn is used
plant the lumps about an inch deep under the surface of
the manure, and about ten inches apart each way. If
the spawn looks very good, and the lumps are large do
not plant them quite so close as when the spawn shows
less mycelium in it, and the lumps are small. Never use
a dibber in planting spawn; simply make a hole in the
manure with the fingers, insert the lump and cover it
over at once, and as soon as the bed has been planted
firm it well all over. Although the lumps are buried
only an inch deep under the manure, we have to make a
hole three or four inches deep to push the lump into to
get it buried.

French or flake spawn is inserted in much the same
way and at about the same distance, only, instead of
cutting it up into lumps, we merely break it into flaky
pieces about three inches long by an inch thick, and in
planting it in the beds, in place of pushing it into the
hole, lay in the flake on its flat side and at once cover it.

Many growers plant spawn a good deal deeper than I
do, but I have never found any advantage in deep planting. In moderately warm beds, or beds that are likely
to retain their heat for a considerable time, I am satisfied that shallow planting is better than deep planting.
When we want to mold over our beds soon after spawning them, shallow planting is to be recommended. But
if the beds are only 75° to 78°, before being spawned;
then I think deep planting is better than shallow planting, because the genial temperature gives the mycelium
a better start in life than would the cooler manure
nearer the surface.

If there is any likelihood of the surface manure getting wet from the condensed moisture of the atmosphere,
I would again cover over the beds with some hay or

straw, and let it remain on until molding time. And
if the bed is a little sluggish,—that is, cool,—this covering will help in keeping it warm. Outside beds should
be molded over in three or four days after spawning;
inside beds in eight to ten days.

Steeped Spawn.—As brick spawn is so hard and
dry I have tried the effect of steeping it in tepid water
before planting; some pieces were merely dipped in the
water, and others allowed to soak in the pails one-half,
one, five, and ten hours. The effect was prejudicial in
every instance and ruinous in the case of the long-soaked
pieces.

Flake Spawn.—"This is produced by breaking up
the brick spawn into pieces about two inches square and
mixing them in a heap of manure that is fermenting
gently. After lying in this heap about three weeks it
will be found one mass of spawn, and just in the right
condition for running vigorously all through the bed in
a very short time. . . . When flake spawn is used
the appearance of the crop is from two to three weeks
earlier than when brick spawn is used."—Mr. Henshaw,
in first edition of "Henderson's Handbook of Plants."
I have tried this method and given it careful attention,
but the results were inferior to those obtained where
plain, common brick spawn had been used at once.

In all my practice I have found that any disturbance
of the spawn when in active growth which would cause
a breaking, exposing, or arresting of the threads of the
mycelium has always had a weakening influence upon
it. I have transplanted pieces of working spawn from
one bed to another, as the French growers do, but am
satisfied that I get better crops and larger mushrooms
from beds spawned with dry spawn than from beds
planted with working spawn from any other beds.

CHAPTER XII.

LOAM FOR THE BEDS.

In growing mushrooms we need loam for casing the beds after they are spawned, topdressing the bearing beds when they first show signs of exhaustion, filling up the cavities in the surface of the beds caused by the removal of the mushroom stumps, and for mixing with manure to form the beds. The selection of soil depends a good deal on what kind of soil we have at hand, or can readily obtain.

The best kind of loam for every purpose in connection with mushroom-growing is rich, fresh, mellow soil, such as florists eagerly seek for potting and other greenhouse purposes. In early fall I get together a pile of fresh sod loam, that is, the top spit from a pasture field, but do not add any manure to it. Of course, while this contains a good deal of grassy sod there is much fine soil among it, and this is what I use for mushrooms. Before using it I break up the sods with a spade or fork, throw aside the very toughest parts of them, and use the finer earthy portion, but always in its rough state, and never sifted. The green, soddy parts that are not too rough are allowed to remain in the soil, for they do no harm whatever, either in arresting the mycelium or checking the mushrooms, and there is no danger that the grass would grow up and smother the mushrooms.

Common loam from an open, well-drained fallow field is good, and, if the soil is naturally rich, excellent for any purpose. But do not take it from the wet parts of the fields. Reject all stones, rough clods, tussocks, and the like. Such loam may be used at once.

Ordinary garden soil is used more frequently than any other sort, and altogether with highly satisfactory results. The greatest objection I have to it is the amount of insects it is apt to contain on account of its often repeated heavy manurings

Roadside dirt, whether loamy or gritty, may also be used with good results. If free from weeds, sticks, stones and rough drift, it may be used at once, but it is much better to stack it in a pile to rot for a few months before using.

Sandy soil, such as occurs in the water-shed drifts along the roads and where it has been washed into the fields, is much inferior to stiffer and more fibrous earth.

I have used the rich dark colored soil from slopes and dry hollows in woods, and, odd though it may appear, as mushrooms do not naturally grow in woods, with success. But it is not as good as loam from the open field.

Peat soil or swamp muck that has been composted for two or three years has failed to give me good returns. The mushrooms will come up through it all right, but they do not take kindly to it.

Heavy, clayey loam is, in one way, excellent, in another, not so good. So long as we can keep it equably moist without making it muddy it is all right, but if we let it get a little too dry it cracks, and in this way breaks the threads of the spawn and ruins the mushrooms that were fed through them.

Loam Containing Old Manure.—Loam in which there is a good deal of old, undecomposed manure, such as the rich soil of our vegetable gardens, is unqualifiedly condemned by some writers because of the quantity of spurious and noxious fungi it is supposed to produce when used in mushroom beds. But I can not join in this denunciation because my experience does not justify it. This earth is the only kind used by many market gardeners, as they have no other, and certainly without

apparent injurious effect. When I was connected with
the London market gardens, some twenty years ago,
Steele, Bagley, Broadbent, and the other large mushroom growers in the Fulham Fields cased all of their
beds with the common garden soil—perhaps the most
manure-filled soil on the face of the earth—and spurious
fungi never troubled them. Indeed, I can not under-
stand why it should produce baneful crops of toadstools
when used in mushroom beds, and no toadstools when
used for other horticultural purposes, as on our carnation benches in greenhouses, in our lettuce or cucumber
beds, or in the case of potted plants. True, spurious
fungi may appear in the earth on our greenhouse benches
or frame beds or mushroom beds at any time and in
more or less quantity, but I am convinced that the rich
earth of the vegetable garden has no more to do with
producing toadstools than has any other good soil, and
old manure has far less to do with it than has fresh
manure.

All practical gardeners know how apt hotbeds, in
spring when their heat is on the decline, are to produce
a number of toadstools; and, also, that when the bed is
"spent," that is, when the heat is altogether gone, the
tendency to bear toadstools has gone too. This peculiar-
ity is more apparent in spring than in fall. All mushroom growers know that spurious fungi, when they
appear at all, are most numerous three to two weeks
before it is time for the mushrooms to come in sight.
The same growth appears in the manure piles out in the
yard; a few weeks after the strong heat of the manure
has gone lots of toadstools may be observed on and about
the heaps, but on the piles of well-rotted cold manure
we seldom find toadstools at all.

The fresh, clean stable manure used in mushroom-
growing is not apt to be charged with the spores of pernicious toadstools; their presence is always most marked
in the case of mixed manures.

And there is a current idea that mushrooms will not thrive in beds in which old manure abounds, either in the loam or fermenting material; that it kills the myce- lium. This, too, I must refute. I have seen heavy crops of spontaneous mushrooms come up in violet and carnation beds in winter, and where the soil consisted of at least one-fourth of rotted manure well mixed with the earth. In cucumber and lettuce beds the same thing has taken place. And in similar beds that have been planted artificially with spawn, good crops of mush- rooms have also been raised, and the mycelium, instead of evading the lumps of old manure in the soil often forms a white web right through them.

CHAPTER XIII.

EARTHING OVER THE BEDS.

This is an important operation in mushroom-growing, and the one for which loam is indispensable. It con- sists in covering the manure beds, after they have been spawned, with a coating, or casing as it is more com- monly called, of loam. The spawn spreads in the ma- nure and rises up into the casing, where most of the young mushrooms develop, and all find a firm foothold. The loam also contributes to their sustenance. And it protects the manure, hence the spawn, from sudden fluctuations of temperature, and preserves it from undue wetting or drying.

The best soil to use for this purpose is rich, fibrous, mellow loam, such as is described, page 100.

If the manure is fresh and in good condition and the beds are in a snug cellar or closed mushroom house, I would not case them until the second week after spawning, say about the eighth or tenth day; but were these same beds in an open, airy shed or other building I would case them over some days earlier, say the fourth or fifth day. A fear is often expressed that when beds are cased within three or four days after being spawned the close exclusion of the manure from the air is apt to raise the heat of the manure in the bed, and thereby destroy the spawn; but I have never known of any truth in this theory, and with well-prepared manure I am satisfied no brisk reheating takes place, at least the thermometer does not indicate it. The great danger of early casing is in killing the spawn by burying it too deep in damp material and before it has begun to run through the manure.

I have conducted several experiments in order to satisfy myself regarding when is the proper time to case the beds, and have found no difference in results between beds that were cased over as soon as they were spawned and others that were not cased over until the fourth, seventh, tenth, or fourteenth day after spawning. The good or bad results in the time of casing depend on the condition of the manure in the beds, the depth at which the spawn has been inserted, the openness or closeness of the place in which the beds are situated, and other cultural conditions. But to delay casing as late as the fifteenth or sixteenth day after spawning is injurious to the crop, because in applying the covering of soil we are sure to break many of the mycelium threads that have by this time so freely permeated the surface of the manure. After the fourth week little white knots may be observed here and there on the spawn threads; these are forming mushrooms, and to delay casing the bed until this time would smother these little pinheads, and greatly mar our prospects of a good crop..

Peter Henderson, in his invaluable work, "Gardening for Profit," has given rise to a deep seated prejudice against molding over mushroom beds as soon as they are spawned by telling us that in his first attempt at mushroom-growing he had labored for two years without being able to produce a single mushroom, and all because he molded over his beds with a two-inch casing of loam just as soon as he had spawned them. Then he changed his tactics, and did not mold over the beds until the tenth or twelfth day after spawning, and was rewarded with good crops of mushrooms. Now, notwithstanding Mr. Henderson's experience, it is a fact that many excellent growers spawn and mold their beds the same day, and with success. But Mr. H. has done much good in displaying a rock against which many might be wrecked, so much depends upon other cultural conditions. The old practice of inserting the spawn three or more inches deep into the manure bed and then molding it at once with two inches deep of loam was enough to destroy the most potent spawn ; nowadays we barely cover the spawn with the manure, and this is how molding over at once is so successful.

All the preparation necessary is to have the loam in medium dry, mellow condition, well broken up with the spade or digging fork, and freed from sticks, stones, big roots, clods, chunks of old manure, and the like.

Sifting the soil for casing the beds is labor lost. Sifted soil has no advantage over unsifted earth, except when it is to be used for topdressing the bearing beds or filling up the holes in their surface.

The condition of the soil should be mellow but inclined to moist. If wet it can only be used clumsily and spread with difficulty ; if dry it can be spread easily but not made firm, and on ridge beds can not be put on evenly. But when moderately moist it can be spread easily and evenly on flat or rounded surfaces, and made firm and smooth.

How deep the mold shall be put upon the bed is also

an unsettled question. Some growers recommend three-fourths of an inch, others one, one and one-half, two, or two and one-half inches, and some of our best growers of fifty or seventy-five years ago were emphatic in asserting three inches as the proper depth, but among recent writers I do not find any who go beyond two and one-half inches. My own experience is in favor of a heavy covering, say one and one-half to two inches. In the case of a thin covering the mushrooms come up all right but their texture is not as solid as it is in the case of a heavy covering, nor do the beds continue as long in bearing; besides, "fogging off" is much more prevalent under thinly covered than under heavily covered beds; also, when the coating of loam is heavy a great many more of the "pinheads" develop into full sized mushrooms than in the case of thinly molded beds.

Opinions differ as to firming the soil. I am in favor of packing the soil quite firm, and have never seen good mushrooms that could not come through a well firmed casing of loam, and I never knew of an instance where firm casing stopped or checked the spreading of the mycelium or the development of the mushrooms. In the case of flat beds,—for instance, those made on shelves and floors,—a slightly compacted coating (and this is all Mr. J. G. Gardner uses) may be all right, but in the case of alongside-of-walls, ridge, and other rounded beds I much prefer and always use solidly compacted casings.

Mr. Henshaw has for several years used green sods about two inches thick, put all over the bed, grass side down, and beaten firmly. The advantage of using sods instead of soil, he thinks, is that the young clusters of mushrooms never damp or 'fogg off' as they are apt to do when soil is used.

I have given this green sods method repeated and careful trials, and am satisfied that it has no advantages, in any way, over common fibrous loam; indeed, it is not as good. No matter how firmly a sod, having its green side down, may be beaten on to a bed of manure, there is barely any union between the two; the sod merely rests upon the dung, but so closely that the mycelium enters it freely. A slight movement or displace ment of the sod after the spawn enters it will break the threads of mycelium between the manure and the sod, and this will destroy the immature mushrooms forming in the sod. This gave me a good deal of trouble. Stepping on the sod would disturb it. A clump of strong mushrooms formed under it sometimes displaces it in forcing their way to the surface.

Sods are only fit for use on flat beds where they can lie solid; on rounded or ridge beds they are too liable to be disturbed. And the trouble and expense of procuring sods are too great to warrant their use, even if they had any advantages. .

CHAPTER XIV.

TOPDRESSING WITH LOAM.

In beds that are in full bearing or a little past their best we often find multitudes of very small or what we call "pinhead" mushrooms, that seem to be sitting right on the top of the loam, or clumps that have been raised a little above the surface by growing in bunches, or what we term "rocks"; now a topdressing of finely sifted fresh loam, about one-fourth to one-half inch thick, spread all over the bed, will help these mushrooms materially without doing any of them harm. But while this topdressing assists all mushrooms that are visible above ground, no matter how small they may be when the dressing is applied, I am not convinced that it induces greater fertility in the spawn, or, in other words, induces the spawn to spread further and produce more mushrooms than it would were no topdressing applied. I know that this is contrary to the opinions and writings of many, at the same time it is according to my own observation.

Go over the bed very carefully and pick out every soft or "fogged-off" mushroom, no matter how small it may be, and root out every bit of old mushroom stem or tough spongy material formed by it, and in this way get the bed thoroughly cleaned. Then fill up all the holes caused by pulling the mushrooms or rooting out the old stumps, and when the whole surface is level apply the topdressing evenly all over the face of the bed, avoiding, as much as possible, burying the well advanced mushrooms. While it would be very well to pack the dressing smoothly over the bed, it is impracticable; we may press it gently with the back of the hand on the bare spots between the mushrooms, but we should not even do this over the mushrooms, no matter how tiny they may be, else many of the "pinheads" will be injured and cause "fogging off."

But we can firm the dressing to the bed by watering it, which may be done over the whole surface of the bed, and without sparing the mushrooms, large or small. Use clear water and apply it gently through a water-pot rose. I always do this, and have never known it to injure the young mushrooms.

In the case of mushroom beds in which black spot has appeared in the crop, I have found that a topdressing of fine, fresh earth applied evenly all over the bed acts, to a certain extent, as a preventive of further attack, but of course has no effect upon any of the already affected mushrooms, large or small.

CHAPTER XV.

The best temperature at which to keep the mushroom house or cellar is 55° to 57°. But much depends upon the method of growing the esculent; the construction of the house or cellar, and other circumstances. Mushrooms can be successfully grown in buildings in which the temperature may be as low as 20° or as high as 65°. By covering the beds well with hay or other protecting material they can be kept warm, even in sharp frosty weather, as the London market gardeners do with their outdoor beds in winter; but when the temperature in the structure in which the mushrooms are grown averages as high as 70° we can not hope for success; indeed, 65° is too high.

A high temperature in a close house or cellar is injurious; it hurries in the crop and forces up the mushrooms weak and thin-fleshed and with ungainly, long stems; it soon exhausts the bed. The time when its evil effects are least visible is early in the fall and late in spring when the outside temperature is high, and when the beds are in somewhat airy rather than close quarters. In the Dosoris cellars there is a steady difference of about 5° in the temperature between the end next the boiler, which is kept at 60° precisely, and that of the farther end, which registers 55° steadily. There is very little difference in the weight of crop produced on the beds at either end of these cellars, but what little there is is in favor of the cooler end. At 60° the crop begins to come in in six to seven weeks after spawning, lasts for three

to four weeks in heavy bearing and a week or more
longer in light bearing, and then it gradually dwindles.

In a temperature of 55° it may be seven weeks after
spawning before the mushrooms appear. In a temperature of 50° they may take a few days longer in appearing, but, as a rule, they are firm, heavy, short-stemmed,
and perhaps a little furry on top and clammy to the
touch, and the beds last in good bearing for two months;
indeed, often a whole winter long. But I have failed to
find that the whole crop from a bed in a 45° to 50° temperature was any greater than that of a like bed in a 55°
to 57° temperature; it is merely a case of getting in six
weeks from the warmer house what it takes ten weeks
to get from the cooler one.

In a temperature of 50° it is not necessary to cover
the beds to increase their warmth, nor is it needful even
in one of 45°, if there is a fair warmth in the body of
the bed to keep the spawn working; but if the warmth
of the interior of the bed falls under 57°, and the atmospheric temperature under 45°, the bed should be kept
warm by covering with hay, straw, matting, or other
material, or better still by boxing it over and laying
this covering on the outside of the boxing. When cold
thicken the covering, when warm lessen it.

CHAPTER XVI.

If the beds get dry they should be watered, for mushrooms will not grow well in dry beds or in a dry atmosphere. Watering is an operation requiring much care. In properly-made beds the manure should remain moist enough from first to last, and whatever dryness is evident should be in the loam casing of the beds and the atmosphere. In all artificially heated mushroom houses the beds and atmosphere are apt to get too dry at one time or another; in underground houses or cellars this is less apparent than in above-ground structures; in shaded north-facing houses dryness is less troublesome than in houses more openly placed.

Endeavor by all fair means to lessen the necessity for watering the beds, but when water is needed never hesitate to give it freely. Mulching the beds and maintaining a moist atmosphere are the best preventives. After the beds are spawned and molded it is a good plan to cover them with a light coating of strawy litter or hay to prevent drying, but this mulching should be removed when it is near time for the young mushrooms to appear. A light sprinkling of water over this mulching every few days, but never enough to reach the soil, assists in preserving enough moisture in the bed under the mulch and also in the atmosphere of the house.

Clean, soft water at a temperature of 80° or 90°; a little warmer or a little colder will not hurt, but do not use water higher than 110°, as it might injure the little pinheads, nor lower than the average temperature of the

111

house, as it would chill the bed, and this should always
be avoided.

Use a small or medium-sized watering pot with a long
spout and a fine rose sprinkler. Apply the water in a
gentle shower over the bed, mushrooms and all, but
never use enough to allow it to settle in pools or run off
in little streams. Clean water sprinkled over the mushrooms does not appear to hurt them, but they should
never be touched with manure water, as it stains them.
Just as soon as the surface of the bed shows signs of
dryness give it water, the quantity depending upon the
condition of the bed. Never let a bed get very dry
before watering it. To thoroughly moisten a very dry
bed requires a heavy watering; so much, indeed, that
the sudden change might injuriously affect the young
mushrooms and spawn. Give enough water at a time to
moderately moisten the soil, not to soak it, but never
sufficient to pass through the soil into the manure.
Clean water only should be used until the beds come
into bearing, but after that time manure water may be
employed with advantage; however, this is not at all
all imperative; indeed, excellent crops can be and are
continually being produced without the aid of manure
water at all.

In the case of beds in full bearing, manure water is
beneficial to the crop. Apply it from a small watering
pot with a long narrow spout but no rose, and pour the
liquid on gently over the surface of the bed, running it
freely between the clumps but never touching any of
the mushrooms. For this reason a rose should not be
used.

I have always used manure water for mushrooms more
or less, but during the past two seasons—'87-'88 and
'88-'89—I have experimented with it continuously and
very carefully, using it in some form or other on part of
every bed, and am satisfied that manure water made

from fresh horse droppings is the best, and the dark colored liquid, the drainings from manure piles, is the poorest; in fact, this latter is not as good as plain water, for it seems to have a deadening rather than quickening effect upon the beds. Cow manure and sheep manure make a good liquid manure, but still I prefer the horse manure, and although having given hen and pigeon manure and guano fair tests I am not satisfied that they have benefited the crop, and there is always a risk in their use. Liquid manure made from the contents of the barnyard tank has not done much good, but fresh urine from the horse and cow stables diluted twelve to fifteen times its bulk has given favorable results.

Mushrooms not only bear with impunity but appear to enjoy a stronger liquid manure more than do any other cultivated plants, and I am satisfied that the weak liquids usually recommended for pot and garden plants would be barely more efficacious than plain water for mushrooms.

The manure water that has given me most satisfaction is prepared as follows : Dump two bushels of fresh horse droppings into a forty-five gallon barrel and fill up with water; stir it up well and let it settle over night. Drain off the liquid the next day and add a pound of saltpeter to it. For use, to a pailful of this liquid add a pailful of warm water. Water of about 80° to 90° is best for mushroom beds. Saltpeter is an excellent fertilizer for mushrooms. I use it in two ways, namely : First, powdered and mixed in the soil for casing the beds, at the rate of two ounces of saltpeter to the bushel of earth. Second, dissolved in water at the rate of two ounces of saltpeter to eight gallons of water, and sprinkled over the beds.

Common salt I use as an insecticide and also as a fertilizer, and am satisfied that it proves beneficial in both ways. Sometimes I sprinkle it broadcast on the surface

8

of the beds, always on the bare places, never touching
the mushrooms, and leave it there for a day or two,
then with a fine, gentle sprinkling of water wash it into
the soil. This is to help destroy the anguillulæ. As a
fertilizer only dissolve four ounces of salt in ten gallons
of water, and with this sprinkle the beds.

A too dry atmosphere can be remedied by sprinkling
the floors, walls, or litter coverings on the beds with
water, not heavily or copiously, but gently and only
enough to wet the surfaces; better moisten in this way
frequently than drench the place at any one time. But
I very much dislike sprinkling the beds in order to
moisten the atmosphere. An experienced man can tell
in a moment whether or not the atmosphere of the mushroom house is too dry. The air in the mushroom house
should always feel moist, at the same time not raw or
chilly, and the floor and wall surfaces should present a
slow tendency to dry up, and the earth on the beds
should retain its dark, moist appearance. The least
tendency to dryness should at once be relieved by damping the wall and floor surfaces.

In houses heated by smoke flues, or still more by ordinary stoves and sheet iron pipes, it may be necessary to
dampen the floors and walls once or several times a day
to maintain a sufficiently moist atmosphere, but where
hot water pipes are used and the houses are tight enough
to require but little artificial heat, such frequent sprinkling will not be necessary. In the case of beds in unheated structures the ordinary atmosphere is generally
moist enough.

Manure Steam for Moistening the Atmosphere.
The late James Barnes, of England, a grand old gardener, writing in the London *Garden*, Vol. III, page 486,
describes his method of growing mushrooms sixty years
ago, and says: "In winter a nice moist heat was maintained by placing hot stable manure inside, and often

turning it over." Mr. John G. Gardner, of Jobstown, N. J., is one of Mr. Barnes's old pupils and a most successful mushroom grower, and he now practices this same method of moistening the atmosphere by hot manure steam. See page 21.

In damping the floors of the mushroom house, as well as the beds, I use a medium-sized watering pot and fine rose; but in sprinkling the walls and other parts not readily accessible by the watering pot I use a common garden syringe.

CHAPTER XVII.

GATHERING AND MARKETING MUSHROOMS.

This is an important point in the cultivation of this esculent, and should be attended to with painstaking discretion.

When mushrooms are fit to pick depends upon several conditions; for instance, whether for market or for home use, and if for the latter, whether they are wanted for soups or stews. For fresh and attractive appearance and best appreciation in the market, pick them when they are plump and fresh and just before the frill connecting the cap with the stem breaks apart. The French mushrooms should always be gathered before the frill bursts; the English mushrooms also look best when gathered at this time, but they are admissible if gathered when the frill begins to burst and before the cap has opened out flat. If the mushrooms display a tendency to produce long stems pick them somewhat earlier, soon enough to get them with short shanks, for long stems are disliked in market; so, too, are dark or discolored or old mushrooms of any sort. Sometimes we may not have enough mushrooms ready at one gathering to make it worth while sending them to market, and are tempted to let them stay ungathered until to-morrow, when they have grown larger and many more shall have grown big enough to gather. This should never be done. It will give an unfavored, unequal lot, some big, some little, some old, some young. Far better pick every one the moment it is ready to gather, and keep all safe in a cool place and covered until some more are ready for use, and in this way have a uniform appearing lot of young produce.

Mushrooms for soups should always be gathered before they burst their gills; indeed, they are mostly gathered when in a button state; that is, when they are about the size of marbles. In this condition, when cooked, they retain their white appearance and do not discolor the soup. Immature mushrooms are deficient in flavor.

For home use, for baking, stewing, broiling, or for cooking in any way in which the tenderness of the flesh

and the delicious aroma of the mushrooms are desirable in their finest condition, let the mushrooms attain their full size and burst their frills, as seen in Fig. 24, and gather them before the caps open out flat,

FIG. 24. A PERFECT MUSHROOM.   or the gills lose any of their bright pink color. If you let them get old enough for the gills to turn brown before gathering, the mushrooms will become leathery in texture, and lose in flavor and darken sadly in cooking.

In picking, always pull the mushrooms out by the

root, and never, if practicable to avoid it, cut them over with a knife.   In gathering, take hold of the mushrooms and give them a sharp but gentle twist, pressing them down at the same time, and they generally part from the bed without any trouble; then place them in the baskets, root-end down, so as to keep them perfectly clean and free from grit.   Sometimes when several mushrooms are joined together in one root-stock and it is impossible to remove one without disturbing the whole, cut it over rather than pull it out.   In the case of clumps of young mushrooms, where one can not be pulled out without displacing some of the others also, cut it out rather than pull it.   There is a knack in pulling mushrooms, easily attained by practice.   And even when they come up in thick bunches and it would appear impossible to pull out the full-grown ones without disturbing the others, a practiced hand will give them a twitch and a pull—they often part from the bed by the gentlest touch—and get them out without unfastening any of the multitude of small buttons that may be growing around them.

The advantages of pulling over cutting are several : It benefits the bed.   If we cut over a mushroom and leave its stump in the ground, in a few days decay sets in and a fluffy or spongy substance grows around the old butt, which destroys many of the little mushrooms around it, as well as every thread of mycelium that comes in contact with it.   One should be particular to scoop out these stumps with a knife before this condition takes place, and go over the beds every few days to fill up the holes, made in scooping out the old stumps, with fresh loam.

Pulled mushrooms always keep fresh longer than do those that have been cut.   In the interest of the market grower they have another advantage.   Mushrooms are bought and sold by weight, and as the stems are always

retained to the caps all are weighed together; if part of
the stems had been cut off the weight would have been
reduced, and, in like proportion, the price; but if the
stems are retained entire not only are the mushrooms
benefited, but the weight, and with it the price, is also
increased.

Gathering Field or Wild Mushrooms.—Go in
search of them in the morning before the sunshine gets
warm and they become too open or old. If you wish to
gather and preserve them in their most perfect condition
pull them up by the "roots," carefully remove any soil
from them, and then lay them orderly in the basket, the
root end down; and by spreading a stout sheet of paper
over the layer, another may be arranged above it in the
same way, and so on until the basket is full. But if you
are not so particular and wish them for immediate use,
or for ketchup or drying, the common way of cutting
them off and carrying them home in bulk will answer
well enough.

Marketing Mushrooms.—Most market growers who
live immediately around New York City sell direct, and
deliver their mushrooms to hotels, restaurants, and
fancy fruiterers. But some of them, also most of those
who live at a considerable distance from the city, sell
their mushrooms through commission merchants in New
York; they, in turn, sell in quantities to suit customers.

Mushrooms are sold by the pound, and come into
market in boxes made of strong undressed paper. Some
growers have light wooden boxes made that hold from
one to four pounds of mushrooms each; and these make
convenient and strong packages for shipping by express.
They may be sent singly, or, as is the case with the
paper boxes, several packed together in crates or boxes.
In sending directly to hotels, cheap baskets, holding one
or several pounds—Mr. Gardner's baskets hold twelve
pounds—are often used, but in sending to commission

merchants, who have to deal them out in quantities to suit customers, mushrooms should always be packed in one, two, three or four pound boxes or baskets, preferably one pound. Mushrooms are not like potatoes or apples, that can be handled, remeasured, and repacked without damaging them. Each rehandling will certainly discolor and perhaps break a good many of them, rendering them unsalable, if not worthless.

The utmost care in gathering and packing of mushrooms for shipping is of primary importance. Gather them the moment they are in best condition, no matter whether or not they are to be packed and shipped the same day; never let them blow open before gathering them; and never cut off short stems. Long stems have to be shortened, but not until everything is ready to pack them. With a very soft hair brush dust off any earth that may stick to the cap of the mushroom, and with a harder brush or the back of a knife rub the earth off of the root end of the stem. Then sort the mushrooms,—the big ones by themselves, the middle-sized by themselves, the small or button-sized ones by themselves, and pack each kind by itself. Pack very firmly without bruising, and so as to show the pretty caps to the best advantage. Never pack mushrooms more than two deep without using plenty of soft paper between the layers, and never put a heavy bulk of them into one box or basket. They discolor so easily that, all things considered, about a pound is enough in a box, if we wish them to carry safely and retain their bright, fresh skin without tarnishing.

Mr. Barter, of London, writes me : "The punnets we use for marketing our mushrooms in are the same that are used for strawberries or peaches. These hold just one pound, but it is becoming more customary now to have little boxes made holding from three to five pounds, as these are better for packing in larger cases for long journeys."

CHAPTER XVIII.

There is a wide-spread impression among horticulturists that worn out beds which have ceased to bear may, by means of watering and certain stimulants and warming up again, be so re-invigorated as to start into full bearing, and yield a second and a good crop. I have given this question much painstaking and practical consideration, and have absolutely failed to revive a "dead" bed. I have not been able to do it myself, and any instance of its having been done has never come under my observation. This may appear heresy anent the multitudinous writings to the contrary.

A mushroom bed may keep on bearing in a desultory way for many months, and now and again show spurts of increased fertility; but this is no second crop; it is merely a prolonged dribbling of the first crop. A bed, by reason of cold or dryness, may, as it were, stand still or partially stop bearing, and soon after it is re-moistened, warmed, and otherwise submitted to congenial conditions, will display renewed energy; but this is no second crop; it is merely a spurt of the first crop caused by extra favorable cultural conditions. But to show how vaguely this question which is so much written about is regarded, let me quote from a letter to me by Mr. J. Barter, who grows 21,000 lbs of mushrooms a year for the London market: "You ask me, 'Do you ever get a second crop?' My beds last in bearing, on an average, each three months, and that I reckon to be three crops. But whether it be three or six months, the weight of mushrooms is about the same. As there is in, say a ton of manure, only so much mushroom-producing power, if you force it to produce that weight in two months you are a gainer, as you thereby save in labor ; but when that producing-power is exhausted it will produce no more mushrooms."

A spent mushroom bed is one that has been kept in bearing condition under the most favorable circumstances at our command, and it has borne a good crop, lasted some two months in bearing, and now it has stopped bearing (except in a meagerly, desultory way) because the spawn or mycelium has exhausted itself and is dead. Then, without living spawn in the bed how are we to get mushrooms ? Some bits of mycelium are still alive and yield the desultory few, but every mushroom that they yield is preying on their vitality, and after a time they too shall die and the bed be completely barren, for the mycelium is altogether dead, and without mycelium mushrooms are an impossibility. We can keep mushroom mycelium in active growth the year round, and year after year, providing we never let it bear mushrooms. This is done by taking the mycelium, just before it begins bearing, from one manure bed and plantit in another, and so on from bed to bed. At every fresh transplanting the mycelium exerts itself into renewed growth, for it must become a strong plant before it has strength enough to produce and support a mushroom. Our utmost efforts have never rendered mycelium in a mushroom-bearing condition perennial.

CHAPTER XIX.

INSECT AND OTHER ENEMIES.

The mushroom grower has his full share of insects to contend with, and in order to overcome them one should acquaint himself with them, and know what they are, what they do, whence they came, and how to destroy them. One should study the diseases and mishaps of his crop and endeavor to know their cause. If we know the cause of failing health in plants, even in mushrooms, we can probably stop or devise a remedy for the disease or means to prevent its recurrence, and if we can not benefit the present subject we are forewarned against future attacks. But there is a deal of mysterious trouble in this direction in mushroom-growing. We are likely to know something about the depredations committed by insects or parasitic molds above ground, but I am sure there is a good deal of mischief going on under ground of which we know very little, if anything. The ills to which the mycelium is subject are not at all fully understood.

"Maggots."—This is the common name among practical mushroom growers for the larvæ of a species of fly (Diptera) which from April on through the warm summer months renders mushroom-growing unprofitable. It is unavoidable, and so far has proved invincible. It attacks the mushrooms in deep cellars, above-ground houses, greenhouses, or frames, and is often quite common in early appearing crops in the open fields. We sometimes read that it does not occur in unheated cellars, but this is a mistake, for in our unheated tunnel cellars, where the temperature in April does not exceed 55°, maggots always appear about the end of this month. But it is true that in the case of cool houses and where the beds are covered over with hay or straw maggots do not appear as early in the season as they do in warm houses and open beds. While rigid cleanliness, and care in keeping the house or cellar closed, no doubt have much to do in lessening the trouble, I have never been able to overcome it, and know of no one who has. We simply stop growing mushrooms in summer.

The maggots or larvæ are about three-sixteenths to four-sixteenths of an inch long, white with black head, and appear in all parts of the mushroom, but mostly in the cap and at the base of the stem, and perforate hither and thither leaving behind them a disgusting network of burrows. The tiny buttons, about as soon as they appear at the surface of the ground, are infested, but this does not check their growth, and when they become mushrooms large enough for gathering, unless it be for a dark looking puncture or tracing now and then visible on the outside of the caps and stems, there are but few signs to indicate to the inexperienced eye the presence of maggots. And this is why maggoty mushrooms are so often found exposed for sale in summer. But in large or full-grown mushrooms, and especially the white-skinned varieties, their presence is visible enough. Although very repugnant, however, and utterly unfit for food, maggoty mushrooms are not poisonous.

But all the mushrooms of summer crops are not maggoty, only a large proportion of them; the evil begins in April, and increases as the summer advances, until August, when it decreases, and in October completely stops—at least this is my experience.

A solution of salt, saltpeter, or ammonia sprinkled over the surface of the beds does not, in this case, do any good as an insecticide, pyrethrum powder diffused

through the atmosphere, and tobacco smoke, have been
ineffectual. Burning a lamp set in a basin of water with
a little kerosene floating on the surface is a most doubtful
operation. Multitudes of flies are destroyed by this lamp
trap, but they are the poor little innocent "manure
flies," and the atmosphere of the house is vitiated and
rendered unhealthy for the crop. I have tried these
lamp traps season after season, and never knew of their
doing any good; that is, the maggots seemed just as
numerous in the lamp-trapped cellar as in the other cellar in which no lamp trap had been used.

Regarding this "maggots" question, Mr. J. F. Barter,
of London, writes me: "During the summer months
the outdoor mushrooms get maggoty before they are big
enough to gather, but of course they can be grown in
cool cellars all the year round. . . . I know of no
sure cure for them (the maggots); of course a slight
sprinkling of salt with manure or mold does prevent, to
a certain extent, but it must be used very carefully."
Now my experience is, as I have already said, that it is
impossible to grow mushrooms here in summer, even in
cool cellars, without having them more or less maggoty.
As regards the salt and loam preventive, I have tried it
lightly and heavily, but without any apparent good
effect.

Black Spot.—All mushroom growers are familiar
with this disease, but unless it appears in pronounced
form very little notice is taken of it, even by market
men, for we see spotted mushrooms continually exposed
for sale. It appears as dark brown spots, streaks, or
freckles, on the top of the mushroom caps, and increases
in distinctness and breadth with age. Fig. 25. It is
caused by eel worms (*Anguillulæ*). These minute
creatures enter the mushrooms when the latter are in
their tiniest pin form and before they emerge from the
ground. If a button arises clean it remains clean, if

diseased it continues to be diseased, and it is a fact that if one mushroom in a clump has black spot we usually find that every mushroom in the clump has it. But mushrooms growing from the same bit of spawn and that come up an inch or two away from the spotted ones may be perfectly clean. Black spot has never occurred with me in new beds, and seldom in those in vigorous bearing, but it generally appears in beds that have been in bearing condition for some weeks or are declining. It does not

confine itself to any particular spot or part of the bed, and sometimes it is much more plentiful than at others. Between October and March we have very little black spot, but as the spring opens this disease increases. During

FIG. 25. MUSHROOM AF-
FECTED WITH BLACK SPOT. the winter season, with careful attention, perhaps not so much as one per cent will show black spot, but as the warm weather sets in the per centage increases until in May, when as many as twenty per cent may be affected by it.

Black spot is a disease, however, that can be controlled. Keep everything in and about the mushroom houses rigidly clean, and as soon as a bed has ceased to bear a crop worth picking clear it out, lime-wash the place it occupied, and make up another bed. Carefully observe that no old loam or manure is allowed to accumulate anywhere, or green scum forms upon the boards, paths, or walls; boiling water impregnated with alum poured over the boards, walls, and other scum-covered surfaces, will kill the eel worms, but it should not be allowed to touch the mushroom beds that are in bearing or coming into bearing. Much can be done to protect the bearing beds from the ravages of this pest: In gathering the mushrooms remove every vestige of old stump and fogged-off mushrooms, keep the holes filled up with fresh loam, and when the bed has been in bearing condition for a fortnight sprinkle it over with a solution of salt, and next day topdress with a half-inch coating of finely sifted fresh loam ; firm it to the bed with the back of the hand, for it can not be pressed on with a spade on account of the growing mushrooms.

Is black spot unwholesome? I do not think so. I have never known any ill effects from eating it. The spotted parts are merely flavorless and tasteless. But it is a very disgusting disease, and no one, I am sure, would care to eat eel worms with their mushrooms. Until quite recently I used to regard the black spot as the mark of some parasitic fungus, and, acting under this impression, sent affected mushrooms to Dr. W. G. Farlow, Prof. of Cryptogamic Botany at Harvard University, for his opinion. He wrote me: "I find that the trouble is due to *Anguillulæ,* and I find an abundance of these animals in the brown spots." He advised me to submit them to an expert in "worms." I then sent samples to my kind friend, Mr. William Saunders, of Washington, D. C., who submitted them, for me, to Dr. Thomas Taylor, the microscopist to the U. S. Department of Agriculture, and who replied: "I recommend that you use a sprinkling of scalding water thoroughly over the entire surface of the bed, especially the portion next to the boxing. The scalding water should be applied before the buttons appear, but not penetrate more than one-eighth of an inch below the surface. Anguillulæ abound wherever decaying vegetable matter exists. . . . The green algæ on the outside of flower pots abounds in the anguillulæ."

Manure Flies.—This is the name we give to the little flies (a species of *Sciara*) that appear in large numbers in spring and summer in our mushroom houses, or, indeed, in hotbeds or structures of any sort where manure is used, as well as about the manure heaps in the yard. On account of their habits they are regarded with much ill-favor. They hop about the house and are continually running over the mushrooms, beds, and walls, in the most suspicious manner. But, notwithstanding this, I am inclined to regard them as perfectly harmless so far as injuring the mushroom crop is concerned, except the fact that they soil the mushrooms somewhat by their traveling over them with their muddy feet.

In attempting to get rid of the maggot fly I have destroyed large numbers of these little innocents, but without any apparent diminution in their numbers. Lachanme recommends: "These flies may be destroyed by placing about a number of pans filled with water to which a few drops of oil of turpentine have been added. The flies are attracted by the odor and drown themselves. They may also be caught with a floating light, in which they will burn their wings and fall into the water." I have found that pure buhach powder dusted into the air or burned on a hot shovel in the mushroom house has been more effective in destroying these flies than either the lamp or drowning process.

Slugs.—These are serious pests in the mushroom house, especially in above-ground structures, and they also occur in annoying numbers in cellars. Wherever hay or straw is used in covering the beds, or there is much woodwork about the house, slugs appear to be most numerous. They are very fond of mushrooms and attack them in all stages, from the tiny button just emerging from the ground to the fully developed plant. In the case of the buttons or small mushrooms they usually eat out a piece on the top or side of the cap, and as the mushroom advances in growth these wounds spread open and display an ugly scar or disfigurement. They also bite into the stems. But in the case of fresh, full grown mushrooms they seem to have a particular

liking for the gills, and eat patches out of them here
and there.

"Bullet" or "Shot" Holes.—My attention was
first called to these by Mr. A. H. Withington, of New
Jersey.   They are little holes cut clear through the
mushroom caps, as if perforated by a buckshot, and are
evidently the work of some insect.   He had, before then,
submitted some of these perforated mushrooms to Prof.
S. Lockwood, who sent them to Prof. C. V. Riley for his
opinion.   Prof. Riley replied that : "It is quite likely
that the damage was done by some myriapod, possibly a
Julus, or some of its allies.   Only observation on the
spot will determine this point."   As I never had any
trouble with myriapods attacking mushrooms and had
seen nothing of this "bullet hole" work in our own
beds I was much interested in the question and deter-
mined to look out for it, so I marked off a part of a bed
and left that uncared for.   I soon found out the trouble.
These holes are the work of slugs which I have found
and watched in the act of eating out the holes.   To find
the slugs at work, one has to take his lantern and go out
and look for them at night.   And to find out about
plant parasites—be they fungus, or insect—one has
to let them alone and watch them.   Had we kept up
our unsparing hunt for slugs, probably we should not
yet have known what caused these "bullet holes," for
no slug would have been left alive long enough to eat a
hole through a mushroom cap.

Slugs must be caught and killed.   We can find them
at night by hunting for them by lamp-light; their slimy
track glistens and reveals their presence.   A few small
bits of slate or half rotten boards with a pinch of bran
on them laid here and there about the beds are handy
traps; the slugs gather to eat the bran, hide beneath
the rotten wood, and can then be caught and killed.
Fresh lettuce leaves make a capital trap, but lettuces in

January or February are about as scarce as mushrooms themselves. A dressing of salt is distasteful to slugs, and not injurious to mushrooms. Strong, fresh lime water may be freely sprinkled over woodwork, pathways, walls, or elsewhere where slugs might gather and hide themselves; but this solution should not be used upon the mushroom beds. Rigid cleanliness, however, about the mushroom house, and an ever-alert eye for slugs, should keep them under.

Wood Lice.—These are sure to be more or less abundant in every mushroom house, even in the cellars. They crawl in through doors, ventilators, or other interstices, and are brought in with the manure, and find shelter about the woodwork, manure, or any bits of dry litter that may be around. They attack the pinhead and small button mushrooms by biting out little patches in their tops and sides; and although these patches are small to begin with, the blemish spreads as the mushroom grows, and is an objectionable feature. Trapping and killing the insects is the chief remedy. Put part of a half boiled potato (for which no salt had been used) into a little pasteboard box, and cover the potato with some very dry swamp moss, lay the box on its side, and open at the end on the bed. The wood lice will gather to eat the potato, and remain after feasting because the dry moss affords them a cozy hiding place. Several of these little boxes can be used. Go through the house in the morning, lift the little traps quickly, and shake out any wood lice that may be in them into a tin pail (an old lard pail will do), which should contain a little water and kerosene. These traps may be used for any length of time, merely observing to change the potato now and again to have it in appetizing condition. Hot water or strong kerosene emulsion may be poured about the woodwork, walls, and pathways, to destroy the wood lice, but should not be allowed to touch the beds. Poisoned sweet apples, potatoes, and parsnips have been recommended as baits for these pests, but I must discourage using poisons of any sort in the mushroom house. Six or eight inch square pieces of half rotten very dry boards laid in pairs, one above the other, also make capital traps; the wood lice gather there to hide themselves; these traps should be examined frequently and the insects shaken into the pail containing water and kerosene.

Mites.—Two kinds of mites are very common about mushrooms in spring and summer; one is whitish and smaller than a "red spider" (one of the commonest insect pests among garden plants), and the other is yellowish and as large as or larger than a "red spider." But I do not think that either of these mites is worth considering as a mushroom pest. The yellow mite (probably *Lyroglyphus infestans*) is extremely common in strawy litter on the surface of hotbeds, and I have no doubt finds its way into the mushroom house as manure vermin rather thàn a mushroom parasite. They are the effect and not the cause of injury to the crop. When mushrooms are wounded or cracked, particularly about the stem, the crevices often become abundantly inhabited with these mites, but they do no material damage.

Mice and Rats.—These rodents are very fond of mushrooms, and where they have access to the beds are troublesome and destructive. Both the common house mouse and the white-bellied fence mouse are mushroom destroyers, but, so far, the nimble but timid field mouse (among garden, open air, and frame crops generally) has never yet troubled our mushrooms, but I can not believe that this immunity is voluntary on its part. The mice bite a little piece here and there out of the caps of the young mushrooms, and these bite-marks, as the mushrooms advance in growth, spread open and become unsightly disfigurements. In the case of open mushrooms, however, the mice, like slugs, prefer the gills to the fleshy caps. Rats are far more destructive than mice. Trapping is the only remedy I use, and would not use poison in the mushroom houses for these creatures for obvious reasons. But we should make our houses secure against their inroads.

Toads.—These are recommended as good insect traps to be used in mushroom houses, but I do not want them there; the cure is as bad as the disease. The mushroom bed is a little paradise for the toad. He gets upon it and burrows or elbows out a snug little hole for himself wherever he wishes, and many of them, too, and cares nothing about whether, in his efforts to make himself comfortable, he has heaved out the finest clumps of young mushrooms in the beds.

Fogging Off.—This is one of the commonest ailments peculiar to cultivated mushrooms. It consists in the softening, shriveling, and perishing of part of the young mushrooms, which also usually assume a brownish color. These withered mushrooms do not occur singly here and there over the face of the bed, but in patches; generally all or nearly all of the very small mushrooms in a clump will turn brown and soft, and there is no help for them; they never will recover their plumpness. Some writers attribute fogging off to unfavorable atmospheric conditions,—the temperature may be too cold, or too hot, or the atmosphere too moist, or too dry. I am convinced that fogging off is due to the destruction of the mycelium threads that supported these mushrooms; it is a disease of the "root," to use this expression; the "roots" having been killed, the tops must necessarily perish. If it were caused by unfavorable conditions above ground we should expect all of the crop to be more or less injuriously affected; but this does not occur; the mushrooms in one clump may be withered, and contiguous clumps perfectly healthy.

Anything that will kill the spawn or mycelium threads

will cause fogging off to overtake every little mushroom
that had been attached to these mycelium threads.
Keeping the bed or part of it continuously wet or dry
will cause fogging off, so will drip; watering with very
cold water is also said to cause it, but this I have not
found to be the case. Unfastening the ground by ab-
ruptly pulling up the large mushrooms will destroy
many of the small mushrooms and pinheads attached to
the same clump; and when large mushrooms push up
through the soil and displace some of the earth, all the
small mushrooms so displaced will probably waste away,
as the threads of mycelium to which they were attached
for support have been severed. A common reason of
fogging off is caused by cutting off the mushrooms in
gathering them and leaving the stumps in the ground;
in a few days' time these stumps develop a white fluff
or flecky substance, which seems to poison every thread
of mycelium leading to it, and all the mushrooms, present and to come, that are attached to this arrested web
of mycelium are affected by the poison of the decaying
old mushroom stump, and fogg off. Any impure matter
in the bed with which the mycelium comes in contact will destroy the spawn and fogg off the young
mushrooms. Lachaume complains about the larvæ of
two beetles, namely *Aphodius fimetarius* and *Dermestes
tessellatus*, which " cause great damage by eating the
spawn, thereby breaking up the reproductive filaments."
Damage of this sort by these or any other insect vermin
will cause fogging off. But I have not noticed either of
the above beetles or their larvæ about our beds.

Flock.—This is the worst of all mushroom diseases
and common wherever mushrooms are grown artificially.
It is not a new disease; I have known it for twenty-five
years, and it was as common then as it is now, and practical gardeners have always called it *Flock*. I say
" worst of all diseases " because *I know* that mushrooms

affected by it are both unwholesome and indigestible, and I can readily believe that in aggravated cases they are poisonous. It is caused by other fungi which infest the gills and frills of the mushrooms, and render them a

hard, flocky mass; sometimes the affected mushrooms preserve their white skin, color, and normal form, at other times the cap becomes more or less distorted. The illustration, Fig. 26, is from life, and a good average of a flock-infested mushroom.

FIG. 26. A FLOCK-DISEASED MUSHROOM.

In gathering mushrooms the growers should insist that every flock-infested mushroom be discarded, and consumers of mushrooms should familiarize themselves with this disease so as to know and reject every mushroom showing a trace of it.

Flock does not affect all the mushrooms in a bed at any time, and I do not believe it spreads in the bed, or, to use the expression, becomes contagious. If one spot of mildew appears upon a cucumber, rose, or grape vine indoors, and is not checked, it soon becomes general all over the plant or plants, and if one spot of mold occurs in a propagating bed and is not checked at once it soon spreads over a large space and destroys every cutting or seedling within its reach, but this is not the case with flock in a mushroom bed. If one mushroom is affected with flock every mushroom produced from that piece of spawn is affected, but not one mushroom produced from the pieces of spawn inserted next to this one is affected by it; not even if the mycelium from the several lumps of spawn forms an interlacing web. If the flock is confined to the mushrooms produced from a certain bit of spawn some may ask, will the other pieces of spawn broken from the same brick produce flock-infested mushrooms? No. I have given this point particular attention, have kept the pieces of each brick close together, and where flock has appeared I have failed to find that the other pieces of spawn from that brick are more liable to produce flock-infested mushrooms than are the pieces of the bricks that, as yet, have not shown any sign of diseased produce.

How general is this disease? In a bed say three feet wide by thirty feet long and of two months' bearing one may get as few as five or as many as fifty flocky mushrooms; one or two may occur to-day, and we may not find another for a week or two, when we may get a whole clump of them, and so on. It is not the large number of them that makes them dangerous, for they never appear in quantity. They sometimes appear among the earliest mushrooms in the bed, but generally not until after the bed has been in bearing condition for a week or two.

What conditions are most favorable or unfavorable to the growth of this disease I do not know; but it is certainly not caused by debility in the mushroom itself, as the parasite attacks healthy, robust mushrooms and debilitated ones indiscriminately. This flocky condition is caused by one or more saprophytic and parasitic fungi of lowly origin, whose various parts are reduced to mere threads, simple or branched, and divided into tubular cells at intervals, or else they are long, continuous microscopic tubes without any partitions, except at those occasional points where a branch, destined to produce spores, is given off. Generally two or more species of these thread-fungi are present at the same time on the mushroom host, and by the multiplied crossing and interweaving of their threads and branches produce, through their great numbers, the whitish, felted mass of "flock"; while as individuals the threads are so minute as to be scarcely or not at all visible to the naked eye. Similar thread-fungi may often be found in the woods among damp leaves, under rotten logs, and on those porous fungi which project, shelf-like, from the trunks of trees. At present there is no way known for destroying the "flock," except to take up and destroy every clump of mushrooms attacked by it. Fortunately the disease is not very serious if proper precautions are observed; for, in our own cellars, where mushrooms have been grown year after year for the past eleven years, we get but few flocky mushrooms in any bed's bearing. The disease is not more common to-day than it was in any former year. But we give our cellars a thorough cleaning every summer.

Cleaning the Mushroom Houses.—After the season's cropping is finished the mushroom houses and cellars should be thoroughly cleaned. Clear out the old beds, and bring outside all the movable floor and shelf boards, scrape up every bit of loose litter or dirt in the place and throw it out, broom down the walls and whatever boarding is left. Whitewash the walls with hot lime wash, and paint every bit of woodwork liberally with crude oil or kerosene. This is to destroy anguillutæ and other insect and fungus parasites. If you wish to use again the boards brought outside, broom them over and paint them copiously with kerosene. And if your cellar or house has a dirt floor, a heavy sprinkling of very caustic lime water all over it will do good in ridding it of vermin.

CHAPTER XX.

In the preface to *Kitchen and Market Gardening*
(London) is the following :

"Mr. W. Falconer and Mr. C. W. Shaw made, in
connection with the London *Garden,* what we believe
to be the first attempt at long and systematic observation of the best culture as it is in London market gardens." This is mentioned to indicate that the writer
speaks on this subject from experience. And although
it is now seventeen years since I became disconnected with
the London market gardens, by revisiting them a few
years ago, and by correspondence and the horticultural
press, I have endeavored to keep informed of all changes
of methods and improvements in culture as practiced
there. At that time Steele, Bagley, Broadbent, Dancer,
Pocock and Myatt were among the largest and best gardeners around London, and since then several of these
grand old gentlemen have passed away and their fields
have been cut up and built upon. At that time mushrooms were one of the general crops, as were snap beans
or cauliflower, and in their season were planted as a
matter of course. To-day they have become a specialty,
and some gardeners devote their whole energy to mushroom-growing alone, and make from $2000 to $5000 a
year clear profit from one acre of mushrooms, and that,
too, from ridges in the open field ! There is no other
field crop that yields such a large profit. There they
get twenty-four to forty-eight cents a pound for their

fresh mushrooms, here we get fifty cents to a dollar a pound for ours. But as mushroom-growing there is confined to fall, winter and spring, those gardeners who restrict themselves to mushrooms only devote the summer months to making mushroom spawn for their own use, and also for sale.

Mr. John F. Barter, of Lancefield street, London, the king of London mushroom growers, writes me under date of Dec. 10, 1888: "I employ men for mushroom bed-making from August to March; then, in order to keep on the same staff, I get about 10,000 bushels of brick spawn made up for sale. . . . By the sale of spawn I make just half of my living." Now let us see: 10,000 bushels = 160,000 bricks, and each brick weighs a pound, thus we have 160,000 pounds. At ten cents a pound (retail price) the total is $16,000; at five cents a pound (supposed wholesale price) $8000, or at three and a half cents a pound (supposed manufacturer's price) $5600.

The manure is obtained from the city stables and hauled home by the gardeners on their return trips from market. The manure collected after midsummer is used for mushrooms, and an effort is made to save the very best horse manure for this purpose. When enough has accumulated for a bed the manure is turned and well shaken, removing only the rougher part of the straw, and thrown into a large pyramidal pile to heat; this shape is adopted as being better than the flat form for keeping out rain. In three or four days the manure is again turned, shaken out and piled up as before; after this it is turned every second day, unless it rains, until it has been turned six or seven times in all. It should then be ready for making into ridges.

The site for the beds should be a warm, well-sheltered piece of ground, either in the open field or orchard; much pains should be exercised to protect it from cold winds. Although a great many mushroom ridges are made under the partial shade of apple and pear trees, I always preferred making them in the open ground. The land should be dry and of a slightly elevated or sloping nature, so that no pools of water can possibly collect on the surface. Having the ground cleared, leveled, and ready, mark it off into strips two feet wide and six feet wide alternately. The two feet wide space is for the mushroom bed, the six feet wide one for the space between the beds; but after the ridges are built, earthed over and covered with straw, they are almost six feet wide at the base. The common sizes of ridges are two feet wide by two feet high, and two and one-half feet wide by two and one-half feet high, and taper to six or eight inches wide at top.

The manure being ready and the site for the beds lined off, the manure is carted to the place and wheeled upon the beds. In making the bed shake out the manure well and evenly to cause it to hold together, tamp it with the back of the fork as you go along, and two or three times before the ridges are completed walk upon and tread the manure down solidly with the feet, and trim down the sides to turn the rain water. Two days after the bed is made up some holes should be bored from the top to nearly the bottom with a small iron bar to let the heat off and prevent the inside of the bed from becoming too dry. Make them about nine inches apart all along the center of the bed. The old gardeners did not use the crowbar. They were very particular not to build their ridges before the chances of overheating were considered past; but notwithstanding all their care some of their beds would get overwarm, when, without a moment's hesitation, they tossed them over, part to the right and part to the left, and left the manure thus exposed for a day or two to cool, and then make up the beds again on the same site.

Brick spawn is always used. Some of those who make a specialty of mushrooms also make spawn for sale as well as for their own use; but the majority of the gardeners prefer to buy rather than make their own spawn.

When the heat has fallen to between 80° and 90° the ridges are spawned, the pieces inserted in three rows along each side, leaving about nine inches between the pieces. A dibber should not be used on any account. The spawn is put in tightly with the hand and the manure pressed down. It should be put in level with the face of the bed, so that the mold may just touch it when the bed is cased. In the event of cold or wet weather, just as soon as the beds are spawned a slight covering of rank litter is laid over them. After a few days this is removed and the beds are molded over with mold from ground to which manure has not been applied for some time. But the general market gardeners do not make this distinction; they use the earth from between the ridges, which has been manured regularly every year for a couple of hundred years or more. The mold is put on evenly with the spade and is about two inches thick at the base of the ridge and one inch thick at top, and well firmed by beating with the back of the spade; indeed, the ridges are now commonly watered through a water-pot rose, again beaten very firmly and the surface left smooth and even. This smooth surface readily sheds rain water, but I question if it has any advantage over a well-firmed unglazed surface. After molding the beds are covered with litter, that is, the rankest straw that had been shaken out of the manure, to a depth of four, six, eight, or ten inches, according to the state of the bed and weather; if the bed is inclined to be cool or if the weather is cold, thicken the covering.

Drenching or long drizzling rains are more injurious to the beds than is cold, and in order to ward them off old Russia mats and any other sort of cloth or carpet covering obtainable is laid over the litter on the beds and weighted down with poles, boards, stones, or anything else that is convenient. Do not disturb this covering for about four weeks, and then on a dry day strip it off and shake up the litter loosely so as to dry it. If there is any white mold on the surface of the soil take a

FIG. 27. THE COVERED RIDGES.

handful of straw and rub it off. If the bed is rather cold put a layer of clean, dry hay next the bed, and on top of this replace the littery covering.

The first beds are made in August, and one or more every month after till March, just as time, convenience and material permit. Summer beds are not attempted unless in exceptional cases. The bulk of the beds are generally put in in September and October. In early fall, also in spring, beds yield mushrooms in about six weeks after spawning; in winter they take eight or nine weeks or more, much depending on the weather.

In cold weather the mushrooms are gathered at noonday; if the weather is windy and it is possible to postpone gathering for another day this is done, as the litter can not be replaced satisfactorily in windy weather. In gathering the mushrooms one man carefully pulls the straw down from the top of the bed, rolling it toward him ; another gathers the mushrooms (pulling them out by the roots, never cutting them) into baskets, and a third man covers up the bed. In this way the three men go up one side of the ridge and down the other, and the work is done expeditiously and well, without exposing any part of the bed more than a minute or two at a time. It is necessary that the uncovering be done by rolling the straw down from the top of the ridge ; if it were rolled up the covering on the other side of the ridge would be sure to slip down a little, and break off many small mushrooms. The mushrooms as gathered are of three grades ; the large or wide-spread ones are called "broilers," the full-sized ones whose neck frill is merely broken about half an inch wide are "cups," and the small white ones whose frills are not broken at all are termed "buttons."   All of these are kept separate. They are marketed in different ways, but the growers who make mushrooms a specialty assort and pack them in chip baskets, boxes, or otherwise, as the metropolitan and provincial markets demand or suggest.   Mr. John F. Barter, writing to me from London, says : "As to punnetts, we use the same as for strawberries or peaches" (the abundance of peaches we have in America is un-known over there), "they hold just one pound." But it is getting more general now to have little boxes made to hold say three to five pounds each ; these are better for packing in larger cases for long journeys."

The first cutting is a light one.   After this the bed is cut twice a week for three weeks in mild weather, or once a week in inclement weather.   The last two or three pickings are thin and only secured once a week. Altogether ten or eleven good pickings are gathered from each bed.

I never knew of a single instance in which any attempt was made to renovate an old or worn-out bed. But when the beds become so dry as to need watering a small handful of salt is dissolved in a large pailful of water and with this solution the beds are freely watered over the straw covering, but never, to my knowledge, under it.

My old friends, George Steele and Mr. Bagley, of Fulham Fields, used to run part of their beds east and west, not only for convenience sake so far as the beds themselves were concerned, but with the view of growing early tomatoes against the south side of these beds in summer, and here they got their finest and earliest crops, for the London gardeners can not grow tomatoes out of doors in the open fields as we can in America. Other gardeners clear away the manure for use elsewhere in their fields, and as it is so well rotted it is in capital condition for cauliflower, lettuces, snap beans, and other crops. But as the mushroom growers who restrict themselves entirely to mushrooms, and who, after the mushroom beds have finished bearing, have no further use for the manure in the spent beds, are always able to dispose of it at one-half the cost price. It is excellent for garden crops and as a topdressing for lawns, on account of its fineness and freedom from all rubbish as sticks, stones, old bottles, old shoes, and the like, is in much demand.

CHAPTER XXI.

In caves and subterranean passages underneath the city of Paris and its environs, thousands of tons of mushrooms are artificially produced every year. These underground caves and tunnels are abandoned quarries from which white building stone and plaster have been excavated, and as the veins of stone permeated through the bowels of the earth, 40 to 125 feet deep, so were they quarried, and the blocks brought to the surface through vertical shafts. It is these tunnels, varying in height and width as the veins of stone varied, that are now used for mushroom-growing. M. Lachaume, in his book, *The Cave Mushroom*, tells us : "In the Department of the Seine there are 3000 quarries ; those which have been abandoned and which are situated close to Paris at Montrouge, Bagneux, Vangirard, Méry, Châtillon, Vitry, Honilles, and St. Denis, are used by the 250 mushroom-growers of the Department. There are several of these quarries with horizontal galleries driven into the calcareous rock from the level of the road, which are mostly large enough to accommodate a good sized cart, but the majority can only be entered, like many coal mines, by vertical shafts 100 to 125 feet deep, down which everything has to pass. The laborers climb up and down a ladder, and the fresh manure is shoveled down the shaft from above, the waste stuff and mushrooms being hauled up in baskets from beneath by means of a windlass."

The manure used is obtained from the Paris stables and furnished by contractors, with whom the mushroom

growers make special bargains because they are very particular about the kind and quality of the manure they
use. Some of these growers use as much as 2000 to
3500 tons of manure each a year for their mushroom
beds. To the caves in the immediate neighborhood of
Paris the manure is hauled out in carts, but to Méry
and other places too far distant to be within easy carting
distance it is sent by rail. The mushroom growers consider that the manure from animals that are worked
hard and abundantly fed on dry, good food is the best;
the droppings from these are always dry and rich in
ammonia, nitrogen and phosphates. The manure from
entire horses that are worked hard they regard as the
best, and, next in value, that from mules. The manure
from horses kept for pleasure, such as carriage and rid-
ing horses, is regarded as poor, notwithstanding the
high feeding of these animals, and the manure from
horses fed on grass or roots, also that of cows, as worthless. Stress is laid on the importance of having a good
deal of urine-soaked straw in the manure, and this is
another reason why manure from draught horses is preferred to that from animals kept for pleasure, as the
bedding of the former is not apt to be kept so clean as
that in aristocratic stables.

The preparation of the manure is conducted near the
mouth of the caves or shafts on a level, dry piece of
ground, and altogether out of doors. As soon as sufficient manure for a pile is obtained it is forked over,
thoroughly shaken up and intermixed, divested of all
extraneous matter such as sticks, stones, bottles, scrap
iron, old shoes, and the like we find in city stable manure, and any dry straw is moistened with water. It is
then squared off into a heap forty inches high and trodden down to thirty inches high. In this state it is left
for about six days, when it is turned, shaken up loosely,
the outside turned to the inside, and all dry parts

watered; the same shallow square form is retained, and it is again trodden down firm. In about six days more it is again turned, shaken up, watered, squared off, and trodden as before. In about three days after this it should be fit for use and may be turned, shaken up loosely, and dumped down the shaft into the cave and carried to the spot where the beds are to be formed. Of course these operations must be modified according to circumstances and the condition of the manure.

In making the beds the ground is first marked off. The first bed is made alongside of the wall, and rounded to the front; the other beds run parallel with this and may be straight, crooked, or wavy, as the interior of the cave may suggest. The beds are all ridge-shaped, eighteen to twenty inches wide at the base, eighteen to twenty inches high in the middle, six inches wide at top, and the sides sloping. Pathways twelve inches wide run between the beds. The workmen build the beds by piece-work and receive one-half cent per running foot. A good workman can make 240 feet a day (*Lachaume*). The beds are built neatly and firmly and with much nicety as regards size and proportions. But the workmen do not use a fork or any other tool in the construction of the beds; they lift, shake up, spread and build the manure with their naked hands and pack it firm with their knees.

The spawn is obtained from the working beds and is what the mushroom growers there call "virgin" spawn, though not at all what we know by that term. As a succession of beds is kept up all the year round it is an easy matter for the growers to get their spawn at any time. The best time to get the spawn is when the young mushrooms are first appearing. A bed or part of a bed in capital working order is selected and broken up and the cakes of manure thoroughly matted up with the active mycelium are selected for spawning the fresh beds.

10

It is asserted that from this active spawn crops of mushrooms appear in twenty days' less time than if dry spawn
were used.

The French spawn is used. Somewhere between the
seventh and fourteenth day after making the bed it will
be in condition for spawning. Break the spawn into
pieces between two and three inches long, two inches
wide, and three-fourths of an inch thick, and insert
these pieces in two rows along the sides of the ridges ;
the first row eight inches above the ground, the second
row eight inches above the first, and the pieces put in
quincunx fashion eight inches apart in the row. The
manure is firmly packed in upon the spawn, the surface
left smooth and even and without being further disturbed
until earthing time.

Much stress is laid upon stratifying the spawn before
using, when dry spawn is employed. About eight days
before a bed is to be spawned the dry spawn is spread
out in a row on the floor of the cave or cellar so that it
may absorb moisture and the mycelium begin to run.
At spawning time these cakes or flakes are broken up
and used in the ordinary way, and, it is claimed, with a
week's difference in favor of the early appearing of the
mushrooms. But no more spawn than is necessary for
immediate use should be stratified, for it will not bear
being dried and damped again.

The chips and powder of the stone which has been
taken out of the quarry and which can be had in abundance on the floor of the quarry or on the surface of the
ground around the shaft, are sifted, and the finer part
saved and mixed with earth in the proportion of three
parts of stone dust to one of earth, and with this the beds
are molded over. The powdered stone is strongly impregnated with salts, so advantageous to the mushrooms.

In seven to nine days after spawning, the beds are
ready for earthing over. This depends upon the condi-

tion of the spawn and how well it has run in the manure.
Before being earthed over the outside surface of the beds
should be covered with white filaments radiating in all
directions which give to the beds a bluish appearance.
When the bed is in the proper state for being covered
with earth the mold is laid on equally and firmly over
the surface about three-fourths of an inch deep. It is
then thoroughly watered through a fine-rosed watering pot
and allowed to settle until the next day, when it is
beaten solid by the back of a wooden shovel. The bed
now needs no further care until the young mushrooms
appear, except a light occasional watering should it
get dry.

In spacious, high-roofed caves the mean temperature
is about 52° F., while in narrow, low-roofed ones it is

FIG. 28. IN THE MUSHROOM CAVES OF PARIS.

about 68°. Of course this makes a wide difference in
the time of bearing and duration of the beds made in
the different caves; those in the warm caves come into
bearing sooner and stop bearing quicker than do those
in the high-roofed caves. On an average the first mush-

rooms appear in about forty days after the beds are spawned, and the beds continue bearing for forty or sixty days, but toward the end of that time the yield diminishes very rapidly.

They are gathered once a day, usually about midnight, so that they may reach the Paris market early in the morning. In size the mushrooms range from three-fourths to one and five-eighths inches in diameter of top, and are pure white in color. The workmen always gather the mushrooms by plucking them out by the roots, and never by cutting them; the gatherers have two baskets, carried knapsack fashion on their back; one is to receive the mushrooms as they are picked, the other contains mold with which to fill in the little holes made by pulling the mushrooms out of the bed. In some caves one man gathers the mushrooms and leaves them in little piles on the bed as he goes along, a woman comes after him and places them in a basket, and a man follows her and fills up the holes with earth. Before bringing the mushrooms up out of the caves they are covered over with a cloth to avoid contact with the outer air, which is apt to turn them brown. They are then placed in baskets that contain twenty-three to twenty-five pounds and sent to market, where they are sold at auction as they arrive. Or they may be sent to preserved-vegetable manufacturers, who contract for them at an all round price.

Proper ventilation is regarded as being of great importance, not only for the sake of the workmen, but also for the mushrooms, which will not thrive in an impure atmosphere. Ventilation is afforded by means of narrow shafts surmounted by tall wooden chimneys whose upper ends are cut at an angle so that the beveled side faces north. In order to avoid sudden changes of temperature and strong draughts, fires, trap doors, and other means employed in assisting the ventilation of coal

FIG. 29. GATHERING MUSHROOMS IN THE PARIS CAVES FOR MARKET.

mines are adopted. To stop strong draughts, too, in
the passages, tall, straw-thatched hurdles are set up.
In narrow caves the breath of the workmen, the gases
given off by fermentation, and the products of combus-
tion of the lamps would soon so vitiate the atmosphere
as to render the caves uninhabitable were they not properly ventilated. Indeed, it frequently occurs that caves
in which mushrooms have been grown continuously for
some years have to be abandoned for a year or two because the crop has ceased to prosper in them. But after
they have been thoroughly cleared of all beds and the
surface soil that would have been likely to be touched
or affected by the manure, and ventilated and rested for
a year or two, mushrooms can again be grown in them
successfully.

CHAPTER XXII.

COOKING MUSHROOMS.

Fresh mushrooms, well cooked and well served, are
one of the most delicious of all vegetables. If we grow
our own mushrooms we can gather them in their finest
form, cook them as we please, and enjoy them in their
most delightful condition. If we are dependent upon
the fields we should be careful to gather only such mushrooms as are young, plump, and fresh, and reject all
that are old or discolored, or betray any signs of the
presence of disease or insects. And in the case of store
mushrooms, that is, the ones we get at the fruiterer's or
other provision store, we should examine them critically
before using them to see that they are perfectly free
from "flock," "black spot," "maggots," or other ail-
ment, and discard all that have any symptoms of disease.

The small, short-stemmed, white-skinned mushrooms offered for sale are of the variety known as French mushrooms, and on account of their white appearance are preferred by many; the longer-stemmed, broader-headed, and darker-colored kind that we also find offered for sale is what is known as the English mushroom. The French mushrooms are the most attractive in appearance and preferred in the market, but the English variety is the best flavored and generally the most liked for home use.

As soon as the frill around the neck breaks apart the mushroom is fit to gather; keeping it longer may add to its size a little, but surely will detract from its tenderness. The gills of the mushrooms will retain their pink tinge for a day after the frill breaks open, but they soon grow browner and blacker, until in a few days they are unfit for food. In gathering, the mushrooms should be pulled and never cut, and kept in this way until ready to prepare them for cooking. By retaining the stem uncut the mushroom holds its freshness and plumpness much longer than it would were the stems removed. Keep them in a cool, dark place, and in an earthenware vessel with a cover or a thick, damp cloth thrown over it; this will preserve their plumpness. If the frill is broken wide apart when the mushrooms are gathered, the caps are apt to open out flat in a day or two, and the gills darken and spread their spores, just as if the mushrooms were still unsevered from the ground.

Carefully inspect the mushrooms before cooking them. If the gills are black and the mushrooms are too old do not use them; if the cap is perforated by insects discard it, as it is very likely there are maggots inside; or if there are dark brown spots ("black spot") on the top of the caps throw the mushrooms away. Old mushrooms are tough, ill-looking, bad-tasting and indigestible, and those infested by insects, although not poison-

ous, are very repugnant, and should not be used. But
the dangerous mushroom is the one affected by "Flock."

Mushrooms should be gathered free from grit; if at
all gritty they require washing, which spoils them. All
large mushrooms should be peeled before they are cooked ;
the skin of the cap parts freely from the flesh, but the
skin of the stem must be rubbed or scraped off. The
gills should not be removed as they are the most delicate meat of the mushroom, but if the mushrooms are
old and intended for soup the gills should be scraped
out with the view of getting rid of their darkening
influence in the soup. In the case of small button
mushrooms, which can not be readily skinned, they
should be rubbed over with a soft cloth dipped in vinegar, so as to remove the outer part of the skin. While
the stems may be retained with the buttons, they should
always be removed from the full-grown mushrooms.

Mushrooms should always be served hot, and they
should be eaten as soon as cooked. In the case of baked
mushrooms and others prepared in a somewhat similar
way they should be covered in the oven by an inverted
dish, soup plate, basin, or the like, and if possible brought
to the table in this way and without the cover removed.
Set the tin upon a mat or cold plate upon the table, then
uncover and serve on hot plates. By this means the
delicious aroma is preserved.

Baked Mushrooms.—Peel and stem the mushrooms,
rub and sprinkle a little salt on the gills, and lay the
mushrooms, gills up, on a shallow baking tin and put a
small piece of butter on each mushroom. Place an
inverted saucer or deep plate over them in the tin, and
put them into a brisk oven for about twenty minutes.
Then take them out and serve upon a hot plate, without
spilling any of the juice that has collected in the middle
of each mushroom. Send to table and eat at once.
This is the common way of cooking mushrooms, and by

it is secured the true mushroom aroma and taste in their perfection.

Stewed Mushrooms.—Peel and stem the mushrooms. Take an enameled saucepan, put a lump of butter in it and melt it, then put in the mushrooms, and season with salt and pepper and a small piece of pounded mace (if you like it), then cover the saucepan tightly and stew the mushrooms gently until they are tender, which will be in about half an hour. Have ready some toast, either dry or fried in butter, as preferred; spread out upon a hot dish, place the mushrooms upon the toast, with the gills uppermost, pour the juice over them, and serve hot. Button mushrooms are the ones usually selected for stewing, but while nicer and whiter they are not so finely flavored as the full sized ones.

Another way of preparing stewed mushrooms is to stem and peel them; dip in water containing lemon juice (this is to prevent their becoming dark-colored in cooking, or giving a dark color to the stew), and drain them dry. Put them into a stewpan, with a good-sized lump of butter and some nice gravy, and let them stew for about ten minutes. Take a little stock or cream, beat up some flour in it quite smooth, and add a little lemon juice and grated nutmeg. Add this to the mushrooms and cook briskly for about ten minutes longer, or until tender.

Soyer's Breakfast Mushrooms.—Place some freshly-made toast, divided, on a dish, and put the mushrooms, stemmed and peeled, gills upward upon it; add a little pepper and salt and put a small bit of butter in the middle of each mushroom. Pour a teaspoonful of cream over each, and add one clove for the whole dish. Put an inverted basin over the whole. Bake for twenty or twenty-five minutes, and do not remove the basin until the dish is brought to the table, so as to preserve the grateful aroma. A delightful dish.

Mushrooms à la Crême.—Peel and stem the mush-rooms, roll a lump of butter in flour and put it into the saucepan, then add the mushrooms and some salt, white pepper, a little sugar and finely chopped parsley. Stew for ten minutes. Take the yolks of two eggs beaten up with two large spoonfuls of cream, and add the mixture gradually to the stew; cook for a few minutes longer, and serve hot. This is a delicious dish, but the fine mushroom flavor is not as pronounced in it as it is in the plain bake or stew.

Curried Mushrooms.—Peel and stem a pound of mushrooms, sprinkle with salt, add a little butter, and stew gently for fifteen or twenty minutes in a little good stock or gravy. Then add four tablespoonfuls of cream and one teaspoonful of good curry powder previously well mixed with two teaspoonsfuls of wheat flour. Mix carefully and cook for five or ten minutes longer, and serve on hot toast on hot plates. A capital dish much enjoyed by those who like curry.

Broiled Mushrooms.—Select large, open, fresh mushrooms, stem and peel them. Put them on the gridiron, stem side down, over a bright but not very hot fire, and cook for three minutes. Then turn them and put a small piece of butter in the middle of each, and broil for about ten minutes longer. Put them in hot plates, gills upward, and place another small piece of butter on each mushroom, together with a little pepper and salt, and flavor with lemon juice or Chili vinegar, and put them into the oven for a minute or two. Then send them to table.

Mushroom Soup.—Take a quantity of fresh young mushrooms, and peel and stem them. Stew them with a little butter, pepper and salt, and some good stock, till tender; take them out and chop them up quite small; prepare a good stock, as for any other soup, and add it to the mushrooms and the liquor they have been stewed in. Boil all together, and serve. If white soup is required use white button mushrooms and a good veal stock, adding a spoonful of cream or a little milk as the color may require. This is a nice soup and tastes good. If the mushrooms are very young they have but little flavor; if they are full grown they darken the soup, and if they arc brown in the gills when used the soup will be disagreeably dark. If, after preparing, but before cooking the mushrooms, you pour some boiling water over them and into this drop a little vinegar or lemon juice, then drain them off through a colander, you can prevent, to a great extent, their darkening influence on the soup, but always at the expense of their flavor.

Mushroom Stems.—The stems of young, fresh mushrooms are excellent to eat, but those of old or stale mushrooms are unfit for food. In the case of plump, fresh, full-sized mushrooms, the upper part of the stem, that is, the portion between the frill and the socket in the cap, is used, but the portion below the frill, that is, the "root" end, is discarded. Any part of the stem that is discolored or tough or woody should be rejected, and only the portion that is succulent and brittle and of a clean white color at any time used. The stems are nearly always retained in "button" mushrooms when they are cooked, and the upper or succulent parts of the stems of plump, fresh, full-grown mushrooms are often cooked along with the caps, but when cooking full-grown mushrooms we prefer, in all cases, to completely remove the stems from the mushrooms, and cook both separately. The stems are not so tender or deliciously flavored as are the caps, but are excellent for ketchup, or flavoring, or a sauce for eating with boiled fowl. In cooking the stems they should be peeled by scraping, for they can not be skinned like the caps.

Potted Mushrooms.—Select nice button or unopen

mushrooms, and to a quart of these add three ounces of
fresh butter, and stew gently in an enameled saucepan,
shaking them frequently to prevent burning. After a
few minutes dust a little finely powdered salt, a little
spice, and a few grains of cayenne over them, and stew
until tender. When cooked turn them into a colander
standing in a basin, and leave them there until cold ;
then press them into small potting-jars, and fill up the
jars with warm clarified butter, and cover with paper
tied down and brushed over with melted suet to exclude
the air. Keep in a cool, dry place. The gravy should
be retained for flavoring other gravies, sauces, etc.

Gilbert's Breakfast Mushrooms. — Get half
grown mushrooms, peel them and lay them, gills-side
upward, on a plate ; put to each a small piece of butter,
but only one layer thick ; pepper and salt to taste ; add
two tablespoonfuls of ketchup and one of water ; press
round the rim of the plate a strip of paste, get another
plate of the same size pressed firmly in the paste ; put
the whole in a brisk oven for twenty-five minutes. The
top plate should be left on until served.

Baked Mushrooms.—(A breakfast, luncheon, or
supper dish.) Ingredients : Sixteen or twenty mushroom flaps, butter, pepper to taste. Mode. For this
mode of cooking the mushroom flaps are better than the
buttons, and should not be too large. Cut off a portion
of stalk, peel the top, and wipe the mushrooms carefully
with a piece of flannel and a little fine salt. Put them
into a tin baking dish, with a very small piece of butter
placed on each mushroom ; sprinkle over a little pepper,
and let them bake for about twenty minutes, or longer
should the mushrooms be very large. Have ready a
very hot dish, pile the mushrooms high in the center,
pour the gravy round, and send them to table quickly
on very hot plates.

Broiled Mushrooms.—(A breakfast, luncheon, or

supper dish.) Ingredients: Mushrooms, pepper and salt to taste, butter, lemon juice. Mode. Cleanse the mushrooms by wiping them with a piece of flannel and a little salt; cut off a portion of the stalk and peel the tops; broil them over a clear fire, turning them once, and arrange them on a very hot dish. Put a small piece of butter on each mushroom, season with pepper and salt and squeeze over them a few drops of lemon juice. Place the dish before the fire, and when the butter is melted serve very hot and quickly. Moderate sized flaps are better suited to this mode of cooking than the buttons; the latter are better in stews.

Mushrooms à la Casse, Tout. — Ingredients: Mushrooms, toast, two ounces of butter, pepper and salt. Mode. Cut a round of bread one-half an inch thick, and toast it nicely; butter both sides and place it in a clean baking sheet or tin; cleanse the mushrooms as in preceding recipe, and place them on the toast, head downwards, lightly pepper and salt them, and place a piece of butter the size of a nut on each mushroom; cover them with a finger glass and let them cook close to the fire for ten or twelve minutes. Slip the toast into a hot dish, but do not remove the glass cover until they are on the table. All the aroma and flavor of the mushrooms are preserved by this method. The name of this excellent recipe need not deter the careful housekeeper from trying it. With moderate care the glass cover will not crack. In winter it should be rinsed in warm water before using.

Stewed Mushrooms.—Ingredients. One pint mushroom buttons, three ounces of fresh butter, white pepper and salt to taste, lemon juice, one teaspoonful of flour, cream or milk, one-fourth teaspoonful of grated nutmeg. Mode. Cut off the ends of the stalks and pare neatly a pint of mushroom buttons; put them into a basin of water with a little lemon juice as they are done. When all are prepared take them from the water with the hands, to avoid the sediment, and put them into a stewpan with the fresh butter, white pepper, salt, and the juice of one-half a lemon ; cover the pan closely and let the musrooms stew gently from twenty to twenty-five minutes, then thicken the butter with the above proportion of flour, add gradually sufficient cream, or cream and milk, to make the sauce of a proper consistency, and put in the grated nutmeg. If the mushrooms are not perfectly tender stew them for five minutes longer, remove every particle of butter which may be floating on the top, and serve.

Broiled Beefsteak and Mushrooms.—Ingredients : Two or three dozen small button mushrooms, one ounce of butter, salt and cayenne to taste, one tablespoonful of mushroom ketchup. Mode. Wipe the mushrooms free from grit with a piece of flannel, and salt ; put them in a stewpan with the butter, seasoning, and ketchup; stir over the fire until the mushrooms are quite done. Have the steak nicely broiled, and pour over. The above is very good with either broiled or stewed steak.

To Preserve Mushrooms.—Ingredients : To each quart of mushrooms allow three ounces of butter, pepper and salt to taste, the juice of one lemon, clarified butter. Mode. Peel the mushrooms, put them into cold water, with a little lemon juice ; take them out and dry them very carefully in a cloth. Put the butter into a stewpan capable of holding the mushrooms; when it is melted add the mushrooms, lemon juice, and a seasoning of pepper and salt; draw them down over a slow fire, and let them remain until their liquor is boiled away and they have become quite dry, but be careful in not allowing them to stick to the bottom of the stewpan. When done put them into pots and pour over the top clarified butter. If wanted for immediate use they will keep good a few days without being covered over. To re-warm them put the mushrooms into a stewpan, strain the butter from them, and they will be ready for use.

Mushroom Powder. — (A valuable addition to sauces and gravies when fresh mushrooms are not obtainable.) Ingredients: One-half peck of large mushrooms, two onions, twelve cloves, one-fourth ounce of pounded mace, two teaspoonfuls of white pepper. Mode. Peel the mushrooms, wipe them perfectly free from grit and dirt, remove the black fur, and reject all those that are at all worm-eaten; put them into a stewpan with the above ingredients, but without water; shake them over a clear fire till all the liquor is dried up, and be careful not to let them burn; arrange them on tins and dry them in a slow oven; pound them to a fine powder, which put into small dry bottles; cork well, seal the corks, and keep it in a dry place. In using this powder, add it to the gravy just before serving, when it will require one boil up. The flavor imparted by this means to the gravy ought to be exceedingly good. This should be made in September, or at the beginning of October, and if the mushroom powder bottle in which it is stored away is not perfectly dry it will speedily deteriorate.

Mushroom Powder. — This is for use as a condiment. The finest full-grown mushrooms — which are the best flavored — should be selected and prepared for drying, and dried as stated under the heading of "Dried Mushrooms," except that it is better to dry them in an oven or drying machine so that they may be dried quickly and become brittle. Grate or otherwise reduce them to a fine powder, and preserve this in tightly-corked bottles.

To Dry Mushrooms. — Wipe them clean, take away the brown part and peel off the skin; lay them on sheets of paper to dry, in a cool oven, when they will shrivel considerably. Keep them in paper bags, which hang in a dry place. When wanted for use put them into cold gravy, bring them gradually to simmer, and it will be found that they will regain nearly their usual size.

Dried Mushrooms.—In the flush of the pasture-mushroom season gather a large number of mushrooms of all sizes and see that they are thoroughly clean ; remove and discard the stems and peel the caps. Stir them around for a few minutes in boiling water to which a little lemon juice or vinegar has been added to prevent them from turning dark colored. Some people use plain cold water, or cold water with lemon juice or vinegar in it. But never use salt in preparing mushrooms for drying, or else the salted mushrooms will absorb moisture from the atmosphere and spoil. Take the mushrooms out of the water and drain them on a sieve, then string them and hang them up to dry and season in an open, airy shed, as one would strings of drying fruit. They may also be dried in a drying machine or oven as one would do with apples or peaches. They are used as a substitute for fresh mushrooms when the latter can not be obtained. In preparing dried mushrooms for use steep them in tepid water or milk until they become quite soft and plump, then drain them dry and cook them in the same way as fresh mushrooms. While they are a good substitute for the fresh article they are deficient in flavor.

Mushroom Ketchup.—To each peck of mushrooms add one-half pound of salt ; to each quart of mushroom liquor one-half ounce of allspice, one-half ounce of ginger, two blades of pounded mace, one-fourth ounce of cayenne.

Choose full-grown mushroom flaps, and be careful that they are perfectly fresh-gathered when the weather is tolerably dry ; for if they are picked during rain the ketchup made from them is liable to get musty, and will not keep long. Put a layer of them in a deep pan,

sprinkle salt over them, then another layer of mushrooms and so on alternately. Let them remain for a
few hours, and break them up with the hand; put them
in a cool place for three days, occasionally stirring and
mashing them well to extract from them as much juice
as possible. Measure the quantity without straining,
and to each quart allow the above proportion of spices,
etc. Put all into a stone jar, cover it up very closely,
put it in a saucepan of boiling water, set it over the
fire and let it boil for three hours. Have ready a
clean stewpan; turn into it the contents of the jar, and
let the whole simmer very gently for half an hour; pour
it into a pitcher, where it should stand in a cool place
until the next day; then pour it off into another pitcher
and strain it into very dry clean bottles, and do not
squeeze the mushrooms. To each pint of ketchup add
a few drops of brandy. Be careful not to shake the
contents, but leave all the sediment behind in the
pitcher; cork well, and either seal or rosin the cork, so
as to exclude the air perfectly. When a very clear,
bright ketchup is wanted the liquor must be strained
through a very fine hair sieve or flannel bag after it has
been very gently poured off; if the operation is not successful it must be repeated until you have quite a clear
liquor. It should be examined occasionally, and if it is
spoiling should be reboiled with a few peppercorns.
Seasonable from the beginning of September to the middle of October, when this ketchup should be made.

Mushroom Ketchup.—This flavoring ingredient, if
genuine and well prepared, is one of the most useful
store sauces to the experienced cook, and no trouble
should be spared in its preparation. Double ketchup is
made by reducing the liquor to half the quantity; for
example, one quart must be boiled down to one pint.
This goes further than ordinary ketchup, as so little is
required to flavor a good quantity of gravy. The sedi-

11

ment may also be bottled for immediate use, and will be found to answer for flavoring thick soups or gravies.

Mushroom Ketchup.—In making ketchup use the very best mushrooms, full grown but young and fresh, as it is highly important to secure fine flavor, and· this we can not get from inferior mushrooms. Take a measure of fine fresh mushrooms and see that they are clean and free from grit; stem and peel them; cut them into very thin slices and place a layer of these on the bottom of a deep dish or tureen; sprinkle this layer with fine salt, then put in another layer and sprinkle with salt as before, and so on until the dish is full. The white succulent part of the stems may also be used in the ketchup, but never any discolored, tough or stringy part. On the top of all strew a layer of fresh walnut rind cut into small pieces. Place the dish in a cool cellar for four or five days, to allow the contents to macerate. When the whole mass has become nearly liquid pass it through a colander. Then boil down the strained liquor to half of its bulk and add its own weight of calf's-foot jelly; season with allspice or white pepper and boil down to the consistence of jelly. Pour into stoneware jars and keep in a cool place.

Pickled Mushrooms.—Use sufficient vinegar to cover the mushrooms; to each quart of mushrooms two blades of pounded mace, one ounce of ground pepper, salt to taste. Choose young button mushrooms for pickling, and rub off the skin with a piece of flannel and salt, and cut off the stalks; if very large take out the red gills and reject the black ones, as they are too old. Put them in a stewpan, sprinkle salt over them, with pounded mace and pepper in the above proportion; shake them well over a clear fire until the liquor flows, and keep them there until it is all dried up again; then add as much vinegar as will cover them; let it simmer for one minute, and store it away in stone jars for use. When cold tie down with bladder and keep in a dry place; they will remain good for a long time. and are generally considered delicious. Make this the same time as ketchup, from the beginning of September to the middle of October. [The above recipes are furnished by Mrs. George Amberley, of New York City.

INDEX.

164

Made in the USA
Monee, IL
07 July 2026